# RATIONALE OF BEINGS

# RATIONALE OF BEINGS

Recent Developments
in
Particle, Nuclear and General Physics

Festschrift in Honor of Gyo Takeda

Edited by **K. Ishikawa, Y. Kawazoe, H. Matsuzaki and K. Takahashi**

World Scientific

*Published by*

**World Scientific Publishing Co Pte Ltd.**
**P. O. Box 128, Farrer Road, Singapore 9128**

Library of Congress Cataloging-in-Publication data is available.

**RATIONALE OF BEINGS**



ISBN 9971-50-117-1

Printed in Singapore by General Printing and Publishing Services Pte. Ltd.

## EDITORS' FOREWORD

On October 13, 1984, the celebration in honor of Professor Gyo Takeda's sixtieth birthday was held. A number of colleagues, friends, former and present students gathered in Aobayama Campus of Tohoku University. There, several talks including the one titled 'Rationale of Beings' by Professor Takeda himself were given.

In order to make the celebration memorable, publication of a festschrift was planned and has been completed in the present form. The title of this book was taken from his talk on that day. The whole volume has been separated into three sections. The first section is devoted to the physics of particles and fields. In the second one the articles on nuclear and atomic physics are presented, and the last section collects papers on general topics. Through this project, we have been able to appreciate how Professor Takeda has done his research and educated his students. We sincerely dedicate this book to him.

We are very grateful to all of the participants. Especially, invited papers are deeply acknowledged. The preface has been written by Professor Robert G. Sachs, from whom Professor Takeda learned the 'American spirit' for physical research during his first stay in U.S.A. as a research associate. We are also indebted to Mitsue Abe, Tadayuki Ohta and Kazunari Shima, who worked together with us for the celebration.

Sendai, Japan, January 1986

Kenzo Ishikawa
Yoshiyuki Kawazoe
Hisao Matsuzaki
Koichi Takahashi

# PREFACE

This volume is composed of articles specially written by associates, former students and other colleagues of Gyo Takeda in celebration of his outstanding qualities as a person, a physicist and an educator on the occasion of his sixtieth birthday. The articles are concerned with today's physics and represent the current broad interests of Takeda and his associates. The evolution of these interests parallels developments in theoretical nuclear and particle physics during the years since Takeda was a student, developments to which he has made important contributions both as a scientist and as a statesman of the physics enterprise in Japan.

Gyo Takeda was a graduate student in theoretical physics at Tokyo University from 1946 to 1950, a most propitious time for beginning a career in field theory, particle physics and high energy physics. The year 1947 alone was the one in which the experimental discovery of the Lamb shift, the pion and the strange particles took place. And it was during the period of Takeda's graduate study that his teacher Sin-itiro Tomonaga along with Schwinger, Feynman and Dyson reported their pioneering work on the renormalization of quantum field theory. Physicists often speak of the discoveries of the neutron, deuteron, positron and neutrino in the year 1932 as the Beginning of the modern era of nuclear, particle and high energy physics. In those terms the period 1946 to 1950 was a Second Beginning and Gyo Takeda is one of the fine products of that era.

As a graduate student, his research consisted of applications of the newly developed methods of quantum electrodynamics to calculations of radiative corrections to collision processes among electrons, protons and photons. He also did some work on pion-nucleon interaction phenomena.

Following his student days, Takeda moved to Kobe University as an Assistant Professor and turned his attention to more general topics. That was the period during which he produced his important contribution to the understanding of renormalization theory in quantum electrodynamics (1952).

When he inquired about the possibility of temporarily joining the theoretical group at the University of Wisconsin, we had already heard of this important contribution and were delighted at the opportunity to take advantage of what we could learn from him. He indicated that by spending a year or two in the United States he hoped to gain experience in relating theory to experiments in particle physics.

He spent three years in the United States, first at the University of Wisconsin and then at Brookhaven National Laboratory. After his arrival in Wisconsin, he quickly demonstrated his determination to meet his proposed goals by collaborating with both the theorists and the experimentalists. For example he learned to scan and interpret tracks in photographic emulsions that had been exposed to high energy particle beams, an important technique of high energy experimental physics in those days.

During his stay at Wisconsin and Brookhaven his theoretical work was primarily concerned with pion-nucleon and pion-nuclear interaction phenomena, dispersion relations and antiproton-proton interactions. At the same time he was a keen observer of the "sociology" and politics of the rapidly developing field of high energy physics in the United States. At that time, the U.S. high energy physicists were trying rather desperately to find a way to adapt to the conflict between scientific individualism and the demands of the Big Science associated with large accelerators. The result was a certain amount of turmoil in the sociology of our science and Takeda had a good opportunity to observe it, to form judgements about the different approaches that were being tried and to apply whatever lessons he found to be appropriate for the development of particle physics and high energy physics in Japan when he returned.

In fact, after his return as Professor in the Institute for Nuclear Studies at Tokyo University he was quickly involved in the planning for accelerators and for experiments. At the same time he began to direct the theoretical work of students and other young theorists who were doing research in strange particle physics. He extended his own theoretical interests to include questions about the meaning of isotopic spin, consequences of parity non-conservation in hyperon decay and electromagnetic interactions of mesons and nucleons.

By 1960 he was well on his way to becoming a statesman of the Japanese high energy physics community. In a letter to me dated in December he comments that he was participating "in too many committees and now the number becomes 20." Apparently he had learned his lessons about the sociology of physics <u>too</u> well! However, he had also learned that one way to solve the problem of too many committees is to take a leave of absence and go to another country where it is possible to find time to study and think while observing the struggles of others over their management problems. It was at that time that he arranged his second extended visit to the United States.

He again went to the University of Wisconsin for some months and then to Lawrence Laboratory in Berkeley, using the opportunity to learn about new developments in theoretical and experimental physics. His theoretical interests turned to studies of Regge poles and S-matrix theory and to phenomenological analysis of pion-nucleon processes. He also renewed his contacts with the community of experimental high energy physicists.

He took up his position at Tohoku University on his return to Japan and, except for a three year interlude as Director of the Institute for Nuclear Studies of Tokyo University, he has been there since then as Professor, also serving two terms as Dean of the Faculty of Science. His second term as Dean ended in May 1985.

The balance in his interests and contributions between fundamental theory and phenomenology and between the science of physics and the general health of the physics enterprise has continued to characterize his activities in Sendai as it had before he went there. He and his collaborators and students have been alert to the shifting winds of particle physics as attested by their continuing research contributions based on new concepts as they have developed, the most recent example being his paper with H. Ui on a Superspace Lagrangian Model of Supersymmetric Quantum Mechanics (1985). The other side of his character, his great sense of responsibility for physics in Japan is demonstrated by his term as Director of the Institute for Nuclear Studies (Tokyo), by his serving as Chairman of the Program Committee for KEK, as a member of the IUPAP Commission on High Energy Physics, as Chairman of the Organizing Committee for the 1978 International Conference on High Energy Physics and as an elected member of many other committees of importance to the Japanese physics and educational establishment.

Further evidence for his enduring interest in new developments in physics is provided by the list of topics included in this volume, representing current interests of those who have been associated with him in one way or another. In addition to those topics mentioned already that relate to his own work in nuclear physics, many particle physics, elementary particle physics, field theory and the foundations of physics, the list includes such topics as information processing, thermally induced broken symmetries, solitons, induced topological term in three dimensions, integral equations for the coulomb problem, phase transitions, fermion fractionization and gravitational potentials for many body systems.

The following comments and acknowledgements taken from some of the articles written for this volume provide an insight into Takeda's role as an educator and leader in physics. Umezawa, Matsumoto and Yamamoto comment on his "original and stimulating arguments based on beautiful intuitions" and present their work on thermally induced broken symmetries in that spirit. At the end of their paper on quasistable solitons, Ezawa and Yanagida acknowledge "many stimulating conversations on the Skyrme model" with Takeda. Ebata, in his paper on vector mesons says: "I recall that my interest in the rho-meson was stimulated by the seminar given by Professor Takeda" when he (Ebata) was a graduate student. Magpantay remarks on Takeda's "concern for the condition of physics and physicists in a country like the Philippines." Akiba says: "My interest in the bag model was stimulated by the lecture of Professor Gyo Takeda." and Iwazaki in his paper on microcanonical quantization expresses appreciation for Takeda's "constant encouragement and educational attitudes." Then there is the acknowledgement by Ohta of his indebtedness for Takeda's "extensive guidance to the world of physics." On a somewhat different note Tsukamoto, Matsuzaki and Kawazoe comment that Takeda thinks "that nowadays everybody depends too much on computers" but they clearly expect him to be tolerant of their work on computer science because of his wide interests.

Gyo Takeda's leadership in physics, in the education of physicists and in science policy is well illustrated by these quotations and by his record of achievement. Because of his strong commitment to physics and his fighting spirit he can be expected to be a source of guidance and inspiration to his students and colleagues in the future as he has been in the past. Above all, in spite of the difficulties imposed by this highly competitive world of physics in which we live, he shows us what it means to be a scholar, a scientist, and a gentleman.

Robert G. Sachs
The Enrico Fermi Institute
The University of Chicago
January, 1986

## ****** CONTENTS ******

### I-4. Fields and Topologies



### I-5. Finite Temperature Field Theory



### I-6. Phenomena of Hadrons



## II. NUCLEI AND ATOMS

## III. GENERAL PHYSICS

# I. PARTICLES AND FIELDS

I-1. Supersymmetry

I-2. Gravity

I-3. Quantum Theories of Fields

I-4. Fields and Topologies

I-5. Finite Temperature Field Theory

I-6. Phenomena of Hadrons

# Supersymmetry and Superconductivity

*Y.Nambu**

Enrico Fermi Institute and Department of Physics
The University of Chicago, Chicago, IL 60637

*ABSTRACT*

The fermionic and bosonic spectra that follow from a BCS mechanism have a rather simple and universal structure. It is shown, in typical examples, that the effective Hamiltonian that reflects such a structure can be factorized in a way similar to the case of supersymmetric systems. The underlying spectrum generating superalgebras are identified.

## 1. Introduction

Recently I made the observation[1] that, in dynamical symmetry breaking of the BCS type, one generally gets simple relations among the gap (mass) parameters of the fermions and the composite bosons, and speculated that such relations are responsible for the apparent supersymmetry[2] in the energy levels of some nuclei. The existence of these relations have been known for a long time, but their universal character has not been fully appreciated. Moreover, it is relatively recently that experiments in superconductors and superfluid helium 3 have confirmed the theoretical results (see ref. 1).

The mass relations I am talking about are as follows. In the case of superconductors with the usual $S$-wave pairing, the masses of the three low energy excitations, i.e., the phase ($\pi$ or Goldstone) collective mode, the quasielectron, and the amplitude ($\sigma$) mode are in the ratios 0:1:2. This is true in the weak coupling limit in which a chain of loop diagrams are summed, and the Ward identity and dispersion relations are used to derive the results. A clear understanding of

* Work supported in part by the NSF, PHY-83-01221, and the DOE, DE FG02-84ER 45144.

the origin of the relations is still lacking, however.

In more complex systems like helium 3, there are a variety of collective modes, but again one can derive a generalized mass formula:

$$m_1^2 + m_2^2 = 4m_f^2, \tag{1}$$

As is already mentioned above, an unanswered theoretical question is what symmetry principle, if any, underlies the mass relations, and in particular whether or not a kind of supersymmetry is inherent in the BCS mechanism. In the present article I shall show that the static (non-kinetic) part of the effective Hamiltonian for the low energy excitations of a BCS system can indeed be expressed as a bilinear form in fermionic composite fields in direct analogy to supersymmetric theories. The system is not supersymmetric, but represents only a slight and almost trivial generalization of supersymmetry. (A preliminary account appears in ref. 3.) In the following sections I will work out the mathematical details, starting from superconductivity and then going to more general cases.

## 2. Fermionic operators in superconductivity.

The low energy excitations in a superconductor can be described by an effective Hamiltonian

$$\bar{H} = \int H dv$$

$$H = \psi^\dagger(\epsilon\tau_3 + f\phi\tau_+ + f\phi^\dagger\tau_-)\psi$$

$$+ (\pi\pi^\dagger + (v^2/3)\nabla\phi \cdot \nabla\phi^\dagger) + f^2(\phi\phi^\dagger - (m/f)^2)^2\,,\ \psi = (\psi_{up}, \psi^\dagger_{dn})\,. \tag{2}$$

Here $\psi$ and $\psi^\dagger$ represent Bogoliubov-Valatin fermionic quasi-particle modes; $\phi$ and $\phi^\dagger$ stand respectively for the bosonic collective modes of electrons and holes; $\pi$ and $\pi^\dagger$ are respectively the canonical conjugates to $\phi^\dagger$ and $\phi$ , $\epsilon$ is the electron kinetic energy measured from the Fermi energy, and $v$ is the Fermi velocity.

The bosonic part of $H$ is of the Ginzburg-Landau form, but with its self-coupling constant dynamically determined to be equal to the square of the Yukawa coupling constant. After the system undergoes the familiar spontaneous breaking of phase invariance (charge conservation), the phase mode $arg$ ($\phi$), the fermionic modes $\psi$ and $\psi^\dagger$, and the amplitude mode $|\phi|$ acquire their masses in the ratios 0:$m$:2$m$.

In the following I will completely ignore the kinetic parts of $H$ for fermions and bosons since they are obviously very different from each other. Now define the fermionic composite fields

$$Q^\dagger = \pi^\dagger\psi_{up} - iV\psi^\dagger_{dn} ,$$
$$Q = \pi\psi^\dagger_{up} + iV\psi_{dn},$$
$$V = f(\phi\phi^\dagger - (m/f)^2) , \tag{3}$$

It is then straightforward to show that their anticommutator gives

$$\{Q(\mathbf{r}), Q^\dagger(\mathbf{r}')\} = \delta(\mathbf{r} - \mathbf{r}')H_s(\mathbf{r}) ,$$
$$H_s = \pi\pi^\dagger + V^2 + \psi_{up}\psi_{dn}\partial V/\partial\phi + \psi^\dagger_{dn}\psi^\dagger_{up}\partial V/\partial\phi^\dagger . \tag{4}$$

$H_s$ is nothing but the static part of $H$. Eq.(4) is to be compared with the case of supersymmetric quantum mechanics:

$$Q = (p - iV(x))\psi , \; Q^\dagger = (p + iV(x))\psi^\dagger ,$$
$$\{Q,Q^\dagger\} = p^2 + V(x)^2 + V(x)' \; [\psi^\dagger , \psi] = 2H. \tag{5}$$

Beyond the rough correspondence $p \to \pi$ , $x \to \phi$, there is a difference in the degrees of freedom of both bosonic and fermionic operators. This difference leads to the consequence that, whereas $Q$ and $Q^\dagger$ commute with $H$ in the case of Eq.(5) because $Q^2 = Q^{\dagger 2} = 0$, these properties are not shared by the operators in Eq.(4). Since spin up and spin down states are not treated symmetrically in $Q$ and $Q^\dagger$, I will also introduce their time-reversed counterparts

$$Q^{\dagger\prime} = - \pi^\dagger\psi_{dn} - iV\psi^\dagger_{up} ,$$
$$Q' = - \pi\psi^\dagger_{dn} + iV\psi_{up} . \tag{6}$$

Each of the four operators in Eqs.(3) and (4) carries a definite charge and spin. The pair $(Q, Q')$ forms a spin doublet with charge -1 (electronlike), and the pair $(Q^{\dagger\prime} , - Q^\dagger)$ forms a doublet with charge +1 (holelike). The anticommutators among them are found to be (omitting spatial delta functions)

$$\{Q,Q\} = -2f\phi S_+ ,$$
$$\{Q' ,Q' \} = 2f\phi S_- ,$$
$$\{Q,Q' \} = -2f\phi S_3 ,$$
$$\{Q,Q^\dagger\} = \{Q' ,Q^\dagger\} = H_s ,$$
$$\{Q,Q^{\dagger\prime} \} = 0 \text{ , etc. ,}$$
$$S_+ = \psi^\dagger_{up}\psi_{dn} ,$$
$$S_- = \psi^\dagger_{dn}\psi_{up} ,$$
$$S_3 = [\psi^\dagger_{up},\psi_{up}]/4 - [\psi^\dagger_{dn},\psi_{dn}]/4 \tag{7}$$

The $S$'s are fermion spin operators. But the algebra does not close here, eventually leading to an infinite tower. On the other hand, if $V$ is linearized from the beginning around the condensates $<\phi> = mz/f$ and $<\phi^\dagger> = mz^*/f$, $|z| = 1$, the resulting four fermionic operators and the four bosonic operators

$$Q_0, Q_0^\dagger, Q_0' , Q_0^{\dagger\prime} \ ;$$
$$\text{and } H_0,\ S_+ ,\ S_- ,\ S_3 \tag{8}$$

form, after appropriate renormalization, a superalgebra $su(2/1)$. In general, an $su(m/n)$ algebra acting on $n$ fermions and $m$ bosons has an even (Lie algebra) part consisting of $su(n)$ , $su(m)$ and a diagonal $u(1)$, the last being weighted in the ratio $m{:}n$ between fermion and boson sectors. (See, for example, ref. 4.) In the present case of $su(2/1)$, the $su(2)$ corresponds to spin, and the $u(1)$ to the Hamiltonian $H_0$, with a mass ratio of 1:2. The zero mass boson does not appear in the algebra because, in the static approximation, it is not a true dynamical degree of freedom. The odd part of $su(2/1)$, i.e., the $Q$'s and $Q^\dagger$'s, act as ladder operators for $H_0$, connecting fermionic and bosonic sectors. Indeed one has

$$[Q_0^\dagger, \bar{H}_0] = -\ mz\ Q_0' \ \ , [Q' , \bar{H}_0] = -\ mz^* Q_0^\dagger \text{ , etc.} \tag{9}$$

Spontaneous charge symmetry breaking thus induces a broken supersymmetry with a supectrum generating superalgebra.

**3. General forms of fermionic operators.**

The previous example suggests that the effective Hamiltonian for a BCS-type system may in general be expressed as a bilinear form in fermionic operators. To see their structure, suppose that the fermion field $\psi$ has $2n$ components forming the fundamental representation of $U(2n)$. (I consider only even numbers because fermions must have half-integral spins, and spin $SU(2)$ is a subgroup of $U(2n)$.) If condensates are formed in $s$ wave pairing, they will belong to the antisymmetric representation

$$\sim \psi_i\psi_j - \psi_j\psi_i \tag{10}$$

and its complex conjugate, each having $k = n(2n-1)$ components.

Let $O_a$, $a = (i,j)$, $i<j$ be $k$ antisymmetric matrices with entries $\pm 1$ respectively at site $(i,j)$ and $(j,i)$, and zero elsewhere. These O's are generators of $SO(2n)$ acting on $\psi$ or $\psi^\dagger$. Correspondingly let $\phi_a$ be $k$ complex scalar matrix fields, and $\pi_a$ be their canonical conjugates.

I now introduce the fermionic fields $Q$ and $Q^\dagger$:

$$Q^\dagger = \pi^\dagger\psi + iV\psi^\dagger,$$
$$Q = \psi^\dagger\pi - i\psi V^\dagger, \tag{11}$$

where the $\phi$'s and $V$ are to be regarded as matrices acting on the $\psi$'s. For simplicity's sake, the coupling constant $f$ is taken equal to 1.

For the "superpotential" $V$, consider the choice

$$V = \phi\phi^\dagger - m^2. \tag{12}$$

The Hamiltonian is generated by the aniticommutators:

$$\bar{H} = \textstyle\sum_n\{\bar{Q}_n, \bar{Q}_n^\dagger\}/2 = \int H dv,$$
$$H = tr(\phi\phi^\dagger\phi\phi^\dagger)/2 - m^2 tr(\phi\phi^\dagger) + nm^4$$
$$+ \psi V'^\dagger\psi/2 + \psi^\dagger V' \psi^\dagger/2\,,\ \ V' = i\textstyle\sum O_a[\pi_a, V] = \phi\,. \tag{13}$$

To see the meaning of this Hamiltonian, first note that $n$ of the $k = n(2n-1)$ fields $\phi_{ij} = -\phi_{ji}$, for example those with

$$i,j = 1,2;3,4;\ldots;2n-1,2n \tag{14}$$

do not directly couple with each other in the potential function in $H$. Keeping only these components, $H$ reduces to

$$H = \sum(\pi_p \pi_p^\dagger + \phi^p \phi_p^\dagger + \psi(\sum O_p \phi_p^\dagger)\psi/2$$
$$- \psi^\dagger(\sum O_p \phi_p)\psi^\dagger/2 \, , \; p = (2i - 1, 2i). \tag{15}$$

Thus the ground state has zero energy, corresponding to the condensates

$$<\phi_p> = z_p m, <\phi_p^\dagger> = z_p^* m \, , \; |z_p| = 1 \, . \tag{16}$$

With the choice of the phase factors $z_p = 1$, the fermion mass term takes the form

$$(m/2)(\psi\eta\psi - \psi^\dagger\eta\psi^\dagger) \, , \; \eta = \sum O_p \, , \; \eta^2 = -1. \tag{17}$$

Thus if the $U(2n)$ symmetry is broken down to $[SU(2)]^n$, each $SU(2)$ subspace is spanned by a complex boson field $\phi_p$ and a pair of fermion fields $\psi_{2i-1}, \psi_{2i}$, and is equivalent to the previous case of superconductivity.

The system after the symmetry breaking, however, has a larger symmetry than $[SU(2)]^n$ because of the equivalence of all the $SU(2)$ subspaces. It is in fact invariant under $Sp(2n)$. There are $2n$ massive fermions and $n(2n - 1)$ massive bosons. From the last example, one would expect the relevant superalgebra to be either $su(2n/[n(2n - 1)])$ or $osp(4n/[2n(2n - 1)])$, the former being a subalgebra of the latter. The Hamiltonian, however, cannot be the $u(1)$ piece of $su(2n/[n(n - 1)])$ for $n > 1$, because the mass ratio does not come out right. I will now proceed to check this out.

Redefine $\phi$ and $\phi^\dagger$ to be the deviations from the condensate given in Eq.( 17) with the standard choice $z = 1$:

$$<\phi> = - <\phi^\dagger> = m\eta \tag{18}$$

(Note that $\dagger$ also implies matrix conjugation.)

Linearize $Q$ and $Q^\dagger$ with respect to $\phi$ and $\phi^\dagger$; call them $Q_0$ and $Q_0^\dagger$, and form $H_0$ out of them. One finds

$$Q_0 = \pi^\dagger\psi + (-\phi + \phi^{\dagger\prime})\eta\psi^\dagger \, ,$$
$$Q_0^\dagger = -\pi\psi^\dagger - \eta(-\phi'^{\,\dagger} + \phi)\psi \, . \tag{19}$$

and

$$H_0 = \{Q_0, Q_0^\dagger\}/2 = -\operatorname{tr}(\pi\eta\pi^{\dagger\prime}\ \eta)/2 - (m^2)\operatorname{tr}(\phi^{\dagger\prime} - \phi)\eta(\phi^{\dagger\prime} - \phi)\eta)/2$$
$$+ m(\psi\eta\psi - \psi^\dagger\eta\psi^\dagger)/2\ ,\ \phi^{\dagger\prime} = \eta\phi^\dagger\eta^{-1}\ ,\ \pi^{\dagger\prime} = \eta\pi^\dagger\eta^{-1}\ . \tag{20}$$

Eq.(19) is written in such a way as to exhibit the invariance of $H_0$ under $Sp(2n)\cap U(2n)$ transformations with respect to the skew metric $\eta$: each contraction of indices is done through the intermediary of $\eta$. In other words, the choice of $V$ given by Eq.(5) corresponds to the case of symmetry breaking $U(2n)$ to $Sp(2n)$. One might suspect that all possible anticommutators of the $4n$ $Q_0$'s and $Q_0^\dagger$'s yield a closed superalgebra, $su(1/2n)$ which has $4n$ odd elements, and contains $sp(2n)$ as a sub algebra. This, however, is not true, as the algebra does not close.

At any rate, these anticommutators are listed in the table. The structure of the algebra for the fermion sector is simple. The $n(2n + 1)$ members in a) are generators of an $sp(2n)$, and together with the $n(2n - 1)$ members in b) form a $u(2n)$ algebra with $(2n)^2$ elements. The fermionic part of $H_0$ is a member of b) with $A = \eta$, and is the $u(1)$ piece of the algebra.

The bosonic part is rather complicated and messy. There are again the same numbers of elements as in the fermionic sector, except for the previous special case of $n = 1$, where the only element was the $H_0$ piece in b). For $n > 1$, however, the algebra in the bosonic sector does not close, due basically to the presence of the factor $\eta$. In fact, the commutators of $Q$ and $Q^\dagger$ with these even elements give rise to new odd elements. Eventually the process closes on a superalgebra $osp(4n/[2n(2n - 1)])$, where the $o(4n)$ is the maximal algebra for the $2n$ $\psi$'s and $2n$ $\psi^\dagger$'s, and similarly the $sp([2n(2n - 1)])$ is the maximal algebra for the $n(2n - 1)$ $\phi$'s and $n(2n - 1)\pi$'s of the massive modes. A difference between $osp(4n/[2n(2n - 1)])$ and its subalgebra $su(n/[n(2n - 1)])$ is that the total fermion number operator (as defined by $\psi^\dagger\psi$) is not in the latter.

The Hamiltonian $H_0$ is invariant under $Sp(2n)$, but it does not occupy a supecial position in the superalgebra except for the case of $n = 1$.

## 4. Comments

The results of the preceding sections can be summarized as follows. The static part of the effective Hamiltonian for the standard superconductor can be expressed as an anticommutator of fermionic composite operators. After symmetry breaking, and in the linear approximation, one gets a spectrum generating superalgebra $su(2/1)$. The Hamiltonian is its $u(1)$ piece with a fermion-boson mass ratio of 1:2. Prescriptions have also been given for a similar procedure for the general case of $2n$ $(n > 1)$ fermions with a symmetry breaking of $U(2n)$ to $Sp(2n)$. The superalgebra is $osp(4n/[2n(2n-1)])$.

A few comments are in order regarding the results of the last section. a) There is quite a bit of arbitrariness in defining the $Q$ operators. Out of $2n$ components of $\psi$ and $n(2n-1)$ components of $\pi^\dagger$ one can construct $2n^2(n-1)$ tensor products, so one must also demand the terms containig the potential function $V$ to have the same degrees of freedom. Together with their Hermitian conjugates, one thus ends up with $4n^2(2n-1)$ composite operators. Now $V$ must be constrained by two requirements: 1) the prescribed pattern of symmetry breaking, and 2) the proper mass relations. The definition of $V$ in Eq.(12) is a deliberately made one for the $U(2n)$ to $Sp(2n)$ breaking and a mass ratio of 1:2. Notice that the order in which $\phi$ and $\phi^\dagger$ is placed in $V$ does not make the $Q$ operators $U(2n)$ covariant. It is covariant only under $SO(n)$, although the resulting $H$ is miraculously $U(2n)$ invariant. A different ordering would have given different mass relations. I do not know, however, whether the solution for the $Q$'s given here is unique or not.

b) A possible physical example of the model systems discussed here is the $SU(4/6)$ nuclear supersymmetry proposed by Iachello and collaborators.[2] In this case, $n = 2$. According to the above results, the superalgebra is $osp(8/12)$. (Its subalgebra $su(4/6)$ in the present sense is not identical with theirs.) The four fermions are the nucleons in the $j = 3/2$ shell, and the six bosons are the collective excitations with $j = 0$ and 2. The Hamiltonian $H$ before symmetry breaking is invariant under $U(4) = U(1) \times SU(4) \sim U(1) \times SO(6)$. The six bosons in the $SO(6)$ interpretation: $\phi_s$, $s = 1,..6$, correspond to

$$\phi_{12} \pm \phi_{34} ,$$

$$\phi_{23} \pm \phi_{14} ,$$
$$\phi_{13} \pm \phi_{24} , \tag{21}$$

in the present $SU(4)$ notation. From this one immediately sees that the condensate matrix $\eta$ in Eq.(17) does represent one of these $SO(6)$ states. The bosonic potential in $H$ may be written in the $SO(6)$ notation as

$$(\textstyle\sum_s \phi_s^\dagger \phi_s - m^2)^2 + \sum_{st}(\phi_s^\dagger \phi_t - \phi_t^\dagger \phi_s)^2/2. \tag{22}$$

Now if the $\phi$'s are expressed as a sum of annihilation and creation operators, and furthermore only the occupation number conserving terms are retained in each factor in Eq.(22), then Eq.(22) is indeed made up of a sum of the Casimir operators of $U(1)$ and $SO(6)$, as would be done in the interacting boson model. However, the fermionic part does not have corresponding potential terms, but only Yukawa coupling terms.

After the symmetry breaking, the residual symmetry of the system is $Sp(4) \sim SO(5)$. There are six Goldstone modes and six massive modes. The fermionic part now becomes an $Sp(4)$ invariant bilinear form. Although this is a suggestive interpretation of the origin of the empirical supersymmetry relations in nuclear physics, some more work would be necessary to substantiate it. In this connection, the second order effects mediated by the Yukawa couplings in $H$ may have to be taken into account.

c) There remain more fundamental questions: Why does the BCS mechanism have a built-in quasi-supersymmetry? Why is it possible to factorize the effective Hamiltonian in the way shown here? Is this always possible? If so, what are the rules for writing down the superpotential $V$ for a given system? Is there a direct derivation of the factorization procedure from the Feyman diagrams that determine the effective Hamiltonian?

I would like to thank P. Freund and J. Rosner for enlightening conversations.

**References**


1. Y. Nambu, Physica **15D** (1985) 147.
2. F. Iachello, Phys. Rev. Lett. **44** (1980) 772; Physica **15D** (1985) 85, and the references cited therein.
3. Y. Nambu, to be published in the Proceedings of the International Conference MESON 50, Kyoto, 1985 (Univ. Chicago preprint EFI 85-71).
4. P. Ramond, Physica **15D** (1985) 25.

| type | fermi sector | bose sector |
|---|---|---|
| a) $\{Q, S\,Q\}$ $= -\{Q', S'\,Q'\}$ | $2i\,\mathrm{tr}(\phi_1\,\pi'_1\,S\,\eta)$ | $[\psi', S\,\eta\,\psi]$ $-[\psi', \eta\,\psi]\,\mathrm{tr}(S)$ |
| b) $\{Q, A\,\eta\,Q'\}$ | $\mathrm{tr}\,\{(\phi_1\,\phi'_1$ $+\pi'_1\,\pi_1)\,S\,\eta\}$ | $\psi'\,A\,\psi'$ $-\psi\,A'\,\psi$ |

$$S^t = S,\quad A^t = -A,\quad S' = -\eta\,S\,\eta,\quad A' = -\eta\,A\,\eta,$$
$$\phi_1 = \phi - \phi'',\quad \pi_1 = (\pi - \pi'')/2,$$
$$\phi'_1 = -\phi_1{}' = \eta\,\phi_1\,\eta,$$
$$\pi'_1 = -\pi_1{}' = \eta\,\pi_1\,\eta.$$

<u>Table.</u> Nonzero anticommutators of Q and Q'.
S and A are arbtrary symmetric and antisymmetric matrices respectively. Only massive bosonic modes $\phi_1$ are retained.

# CONSTRAINING THE STANDARD MODEL WITH A SUPERGROUP

Yuval Ne'eman[+]

Sackler Faculty of Exact Sciences[*]
Tel Aviv University, Tel Aviv, Israel

## DEDICATION

It is a pleasure dedicating this essay to Gyo Takeda upon his Sixtieth Birthday. Aside from learning from his work, I greatly enjoyed collaborating with him twenty years ago at the First Tokyo Summer Lecture Series in Theoretical Physics. My family and I relaxed at that

おいそ　ゴルフ　コス　クルブ

and we even enjoyed the typhoon that followed, thanks also to the help and guidance provided by Y. Hara, H. Miyazawa, N. Nakanishi.

## 1. THE ARBITRARY PARAMETERS OF THE ELCTROWEAK THEORY

The most recent determination[1)] of $\sin^2\theta_W$, by the Chicago-Caltech-Columbia-Fermilab-Rochester-Rockefeller groups, form deep-inelastic scattering of $\nu_\mu$ and $\bar{\nu}_\mu$ on iron nuclei, has yielded a value of .242 ± .011 ± .005. This is rather close to .25, a value obtained from the (super) group[2)] SU(2/1) used as a constraining structure

$$SU(2/1) \supset SU(2)_W \times U(1)_W \tag{1}$$

In the electroweak $SU(2)_W \times U(1)_W$ (or better[3)] $U(2)_W$) this is a free parameter. "Conventional" wisdom assumes that further constraining should come from a GUT

$$G \supset SU(3)_{QCD} \times U(2)_W \tag{2}$$

---

* Wolfson Chair Extraordinary in Theoretical Physics
Supported in part by the U.S.-Israel Binational Science Foundation

\+ Also University of Texas, Austin and supported in part by U.S. DOE Grant DE-FG05-85ER40200

and the most popular[4] $G$ = SU(5) yielded a value of 3/8 for the symmetric case, and spontaneous symmetry breaking (SSB) in the region above $10^{15}$ GeV (constrained by the proton lifetime lower bound) renormalizes[5] this value to $\sim .2$ (the most stretched value might be .22). However, with the proton lifetime definitely ruling out[6] this simple SU(5) model anyhow, there is at present no prediction deriving from $G$. Note that at the same time, the masses of the $W^{\pm}$ and $Z^{o}$ yield the same value for $\sin^2\theta_w$, thus supporting the simplest (one Higgs field) SSB for $U(2)_w$, also embedded in SU(2/1).

For one "generation" of quarks and leptons, the input of $U(2)_w$ includes the selection of 3 multiplets $(|I_w|, Y_w)$ for the quarks $q_L(^1/_2, ^1/_3)$, $u_R(o, ^4/_3)$, $d_R(o, -^2/_3)$, two for the leptons $\ell_L(^1/_2, -1)$, $e_R(o, -2)$, and one for the Higgs field $\phi(^1/_2, 1)$ and its conjugate $\bar{\phi}(^1/_2, -1)$. The vector-meson multiplet is implied in the selection of $U(2)_w$ itself. Altogether, the algebraic input includes eight items (including $\theta_w$). In SU(2/1) we select one multiplet for the quarks; the lepton multiplet then arises automatically. The Higgs fields are also prescribed by the group. We thus make just <u>two algebraic selections altogether</u>, instead of eight! Moreover, even these two can be heuristically derived from first principles.

## 2. THE LIE SUPERALGEBRA $\mathfrak{su}(2/1)$

For definitions, notation and basic results on superalgebras and supergroups, we refer the reader to our exposition in ref.[7]. The $\mathfrak{su}(2/1)$ superalgebra is defined by the graded-bracket

$$[\mu_A, \mu_B\} = 2\, C^E{}_{AB}\, \mu_E \tag{3}$$

where $\mu_A$ are 3x3 supertraceless matrices

$$\mathrm{str}\, \mu_A = o \tag{4}$$

For the $\mathfrak{su}(2/1)$ matrices we use the notation

$a,b,c = 1,2,3;\quad u,v = a,8;\quad i,j,k = 4,5,6,7$

$$\begin{aligned}
&\mu_u \varepsilon L^o \quad , \quad \mu_i \varepsilon L^1 \\
&C^c{}_{ab} = i\, f_{abc} \quad \text{or} \quad -iC^c{}_{ab} = e_{abc} \\
&C^c{}_{a8} = 0 \\
&C^j{}_{ai} = i\, f_{aij} \qquad C^j{}_{8i} = \frac{1}{3}\, f_{8ij} \\
&C^a{}_{ij} = d_{ija} \qquad C^8{}_{ij} = -\frac{\sqrt{3}}{2}\delta_{ij}
\end{aligned} \tag{5}$$

The $f_{Aij}$ and $d_{ijA}$ refer to SU(3) coefficients[8].

Equations (5) define a u(2) subalgebra and subgroup which we identify with $U(2)_w$. $L^1$ behaves as a complex spinor under that $U(2)_w$

$$\begin{aligned}
&C^j{}_{ui} : (146)=(157)=(245)=(256)-(346)-(356)-(347)=(357)=(846)= \\
&\qquad = (847)=(856)=(857) = 0 \\
&-i\, C^j{}_{ui} : (147)=-(156)=(246)=(257)=(345)=-(367) = 1/2 \\
&\qquad (845)=(867)=(2\sqrt{3})^{-1}
\end{aligned} \tag{6a}$$

The $C^j{}_{ui}$ are totally antisymmetric. We define for $\mu^i$

$$|\vec{I}_w| = |C^j{}_{ai}| = 1/2 \quad , \quad |Y_w| = 2\sqrt{3}\,|C^j{}_{8i}| = 1 \tag{7}$$

Equation (5) describes the $\{\mu_i,\mu_j\}$ anticommutators, with

$$C^a{}_{ij} : (146)=(157)=-(247)=(256)=(344)=(355)=(366)=-(377)=1/2 \tag{6b}$$

all components other than (6b) or $C^8{}_{ij}$ of (5) vanishing.

The matrices $\mu_A$ cannot be normalized directly by $\mathrm{str}(\mu_A)^2$ as can readily be seen for the $\mu_i$. These matrices have $\mathrm{str}(\mu_A)^2 = o$ which defeats normalization.

Normalization thus proceeds in the following manner. Under $L^O$, the $\mu_A$ separate into two representations: $L^O$ itself, and L'. One could thus allow at most two normalizations by $\mathrm{tr}(\mu_A)^2$, one for the $\mu^u$ as generators of a connected $U(2)_w$, and one for the $\mu_i$ as the $|I_w| = \frac{1}{2}$, $y_w = \pm 1$ irreps of $U(2)_w$. We pick an overall normalization in this 3-dimensional defining representation (only!)

$$\mathrm{Tr}\,(\mu_A\, \mu_B) = 2\, \delta_{AB} \tag{7}$$

For SU(2/1), we can then define a Killing metric[9],

$$g_{AB} = \text{str}\,(\mu_A\,\mu_B)$$

$$g_{ab} = 2\,\delta_{ab}\quad , \quad g_{88} = -1/3 \tag{8}$$

$$g_{45} = -g_{54} = g_{67} = -g_{76} = 2i$$

Denoting $g^{AB}$ the inverse matrix and using

$$g^{AB}\,\mu_B = \mu^A$$

we get

$$\text{str}(\mu_A\,\mu^B) = \text{str}(\mu_A\,\mu_C\,g^{CB}) = g_{AC}\,g^{CB} = \delta_A{}^B \tag{9}$$

The diagonal (Cartan subalgebra) basis is thus given by

$$\mu_3 = \begin{pmatrix} 1 & & \\ & -1 & \\ & & \end{pmatrix} \qquad \mu_8 = \frac{1}{\sqrt{3}}\begin{pmatrix} -1 & & \\ & -1 & \\ & & -2 \end{pmatrix} \tag{10}$$

and all other $\mu_a$ and $\mu_i$ are identical with the SU(3) $\lambda_a$ and $\lambda_i$.

A ninth u(2/1) base vector $\mu_9$ is trace-orthogonal to $L^o$ but not supertrace orthogonal to su(2/1). It is thus not generated by closure of su(2/1) even though it does not commute with it.

$$\mu_9 = \sqrt{\frac{2}{3}}\begin{pmatrix} 1 & & \\ & 1 & \\ & & -1 \end{pmatrix} \tag{11}$$

Note that the identity is supertrace-orthogonal to su(2/1).

We fix the scale of the charges $M_A$ so that

$$M_A \sim \frac{1}{2}\,\mu_A \quad , \qquad \{M_A, M_B\} = C^E{}_{AB}\,M_E \tag{12}$$

and identify the following physical quantum numbers (L stands for left-chiral, E the electric charge)

$$I_W^a = I_L^a = M_a \quad , \qquad Y_W = 2\sqrt{3}\,M_8, \qquad E = I_L^3 + \frac{1}{2}\,Y_W \tag{13}$$

SU(2/1) admits a discrete Z(3) subgroup, the valency group:

$$v_o = 1$$

$$v_{\pm} = \text{diag.}\ (\exp \pm 2\ \pi i/_3,\ \exp \pm 2\ \pi i/_3,\ \exp \mp 2\ \pi i/_3 \qquad (14)$$

$$= \exp \pi i \sqrt{\tfrac{2}{3}}\, \mu_9$$

Unlike the triality subgroup of SU(3) it does not belong to the center of SU(2/1) which is trivial.

In ref.[10] we have discussed various types of conjugation existing in SU(2/1). In addition we define "star -conjugation"

$$(\mu_A)^{\#} = [2(\mu^A)^{*}]^{sT} \qquad (15)$$

where (*) is complex-conjugation, ($\sim$) is ordinary transposition, and ($^{sT}$) is supertransposition, a rearrangement of the submatrices

$$\left|\begin{array}{c|c} A & \beta \\ \hline \gamma & D \end{array}\right|^{sT} = \left|\begin{array}{c|c} \tilde{A} & -\tilde{\gamma} \\ \hline \tilde{\beta} & \tilde{D} \end{array}\right| \qquad (16)$$

with $A, D \in \Omega^{(+)}$ ; $\beta, \gamma \in \Omega^{(-)}$

the even and odd parts of a Grassmann algebra $\Omega$ for the supergroup. We shall replace the ordinary Hermiticity requirement by star-Hermiticity,

$$(\mu_A)^{\#} = \mu_A \qquad (17)$$

## 3. REPRESENTATIONS AND ASSIGNMENTS

The finite-dimensional representations of $su(2/1)$ have been classified[11]. They are characterized by the quantum numbers of the highest weight $|\hat{y}, \hat{i}_3\rangle$ , which is the state with the highest y and $i_3$ values in a submultiplet with highest $|I_w|$. The most general representation contains four $U(2)_w$ submultiplets:

| | (a) | (b) | (c) | (d) |
|---|---|---|---|---|
| $Y_w$ : | $\hat{y}$ | $\hat{y}+1$ | $\hat{y}-1$ | $\hat{y}$ |
| $\vert I\vert$: | $\hat{i}_3$ | $\vert-\hat{i}_3+\frac{1}{2}\vert$ | $\vert\hat{i}_3-\frac{1}{2}\vert$ | $\vert\hat{i}_3-1\vert$ |

(18)

Example: the adjoint (o,1), to which we have to assign the $U(2)_w$ vector mesons $(W^a_\mu, B^8_\mu)$ for (a,d), and $K^*(890)$-like fermionic ghosts $\beta^i_\mu$ for (b) and (c). In addition, we should assign to the same representations (but with inverted statistics) the spinless fermionic ghost $\alpha^u$ of the Yang-Mills multiplet (in a,d) and the Higgs multiplets $\phi^i$, $\bar{\phi}^i$ (in b,c). We note that the complete description of the representation thus requires an indication of the statistics distribution. We shall denote as "normal" representations with $\nu = +1$ (the "valency") those in which the statistics of the highest weight obeys the rule $(-1)^{2\hat{i}_3} = +1$ for Bose and $-1$ for Fermi statistics and as "abnormal" ($\nu = -1$) those representations where the statistics are inverted. The valency will appear as a $\pm$ sign over $\hat{i}_3$. The vector mesons are thus in $(o,1^-)$.

For $\hat{i}_3 = \frac{1}{2}$, the general case does not generate (d) in (18), and there are just three $U(2)_w$ submultiplets. The quarks are (up to statistics) in $(^1/_3, ^1/_2)$, the antiquarks in $(-^1/_3, ^1/_2)$.

For $\hat{y} = \pm 2\,\hat{i}_3$, the representation further reduces to just two $U(2)_w$ submultiplets: (a) and (b) for the (+) case, (a) and (c) for (-). The leptons and antileptons are assigned (up to statistics) to $(-1,\frac{1}{2})$ and $(1,\frac{1}{2})$ respectively.

The values of the weak hypercharge in the quark representation can be displaced by a constant K,

$$q : (^1/_3 + K,\ ^1/_2) \qquad \bar{q} : (-^1/_3 - K,\ ^1/_2)$$

however, when $^1/_3 + K = n$, an integer, this 4-component representation reduces $\underline{4} = \underline{3} \oplus \underline{1}$. Thus, $\mathfrak{su}(2/1)$ automatically predicts the separation of the right-handed neutrino!

Note that if we base our physical definition of SU(2/1) on the assignment of the Higgs-fields together with the Yang-Mills ghosts in the adjoint representation $(o,1^-)$, we obtain the quark and lepton representations automatically!

Coming back to the statistics of the submultiplets in the matter representations, we notice that $\mu_9$ of equation (11) is proportional to the helicities of the leptons whose $(y, |i|, i_3)$ quantum numbers are given by the above representations. Thus, the grading in $su(2/1)$ coincides with helicity.

The left-handed leptons are, therefore, assigned to $(-1,\frac{1}{2}+)$, the left-handed quarks to $(^1/_3,{}^1/_2{}^+)$; their right-handed antiparticles belong to $(1,\frac{1}{2}^+)$ and $(-^1/_3,{}^1/_2{}^+)$. The right-handed leptons are assigned to $(-1,\frac{1}{2}^-)$, the right-handed quarks to $(^1/_3,{}^1/_2{}^-)$ and their left-handed antiparticles to $(1,\frac{1}{2}^-)$ and $(-^1/_3,{}^1/_2{}^-)$. The states with abnormal statistics in all four multiplets are ghosts; they correspond to inverted statistics (i.e. bosons) with inverted helicities to the leptons or quarks with the corresponding $(y, |i|, i_3)$ quantum numbers.

The value of $\theta_w$ can be read from the defining matrix representation. We find $\tan\theta_w = 1/\sqrt{3}$, so that $\sin^2\theta_w = .25$.

From the symmetry, the mass of the Higgs meson is predicted to be $m_H \sim 250$ GeV. However, this result assumes a Lagrangian constructed with an SU(3) Euclidean metric, a puzzling feature. The same assumption yields $\lambda = \frac{4}{3} g^2$, g the SU(2) gauge coupling. This approach can be further extended to include $SU(3)_{color}$ and the observed sequential pattern of quark/lepton "generations"[12]. A model based on SU(7/1) predicts[13] eight generations, four of them with inverted chiralities. This happens to correspond (16 "flavours") to "critical QCD" and is said to explain uniquely the observed features of high energy scattering[14].

## 4. NEW TYPES OF GHOST FIELDS

The manner in which a supergroup acts internally is not clear a-priori, as the graded vector spaces carrying its action should be bringing together states with identical behaviour under the Lorentz group, but opposite

Quantum statistics. We have introduced in our assignments the conjecture that the $su(2/1)$ superalgebra transitions relate particle fields to ghost fields, of the type introduced by Feynman, DeWitt, Faddeev and Popov to ensure the Unitarity of Quantized Gauge Theories. $su(2/1)$ thus resembles the Becchi-Rouet-Stora[15] and the Curci-Ferrari[16] algebras, also relating particle and ghost fields, and providing algebraic shortcuts, simplifying the imposition of Unitarity throughout the renormalization procedure. However, we have also noted that for the spin ½ (matter) fields, the grading of the carrier vector-spaces is correlated with chirality. Indeed, we have proved elsewhere[17] that the (reducible) matrices representing $U(2)_W \times SU(3)_{color}$, when acting on supermultiplets constructed from juxtapositions of independent matter multiplets and graded according to chirality (thus disregarding Lorentz invariance in that auxiliary construction) have vanishing supertraces. It is thus not surprising that they should in that form be embeddable in Unitary Superunimodular groups such as SU(2/1) or SU(7/1), whose generator superalgebras are supertraceless by definition. However, we have to replace the particles in either one of the gradings by ghosts (to ensure a change of statistics under the action of the odd part of the superalgebra) with inverted chiralities. For example, the two submultiplets of e-leptons $({}_e\nu^o_L, e^-_L)$ with $I_W = ½$, $Y_W = -1$ and $(e^-_R)$ with $I_W = 0$, $Y_W = -2$ can be combined as a reducible $({}_e\nu^o{}_L, e^-{}_L)_1 \oplus (e^-{}_R)_o$, the external index defining a grading (as yet without the action of a superalgebra). The matrices of $su(2)_I$ act only on the odd-graded left-chiral doublet, and are traceless; the $u(1)_W$ matrix has equal traces -2 in both gradings. The difference between the traces of the two gradings thus vanishes for the entire $su(2) \times u(1)$. However, when these matrices are embedded in $su(2/1)$, the superalgebra's (irreducible) 3-dimensional carrier space $(-1, ½^+)$ has to be $({}_e\nu^o_L, e^-_L / \chi_L(e^-_R))$, where

$\chi_L(e^-_R)$ is a left-chiral "matter" ghost field with the quantum numbers of the right-chiral electron $e^-_R$. The dynamical role of these previously unencountered $\chi_L(e^-_R)$ matter ghosts was clarified in a study[18] in which we constructed the relevant Curci-Ferrari algebra for the SU(2/1) gauge and matter multiplets and showed that the ghost-antighost doubling occurring in the Curci-Ferrari construction ensures that for each state in the matter multiplet one gets a total of $|1-1-1| = 1$ physical degrees of freedom (the ghost and antighost count in the negative).

In the 8-dimensional adjoint representation $(0,1^-)$ with spin J=o, a most satisfactory feature is the appearance of the four $U(2)_W$ Faddeev-Popov ghosts $\alpha^u$ together with the Goldstone-Higgs field $h^i$. However, for J=1, the intermediate bosons (and electromagnetic field) $A_\mu$ are accompanied by yet another previously unencountered vector-ghost field $\beta^i_\mu$. The conventional Weinberg-Salam Lagrangian (with spontaneous symmetry breakdown) was shown by 't Hooft and others to produce a unitary theory without such $\beta^i_\mu$ ghosts. Is there a version of the theory requiring it?

## 5. "EXTERNAL" GAUGING OF AN "INTERNAL" SUPERSYMMETRY

In ref.[19] we have suggested a method for the gauging of an "internal" supergroup. The exponentiation of a superalgebra $g = g^+ + g^-$ (the even and odd parts) requires[10] the parameters to belong to the even and odd parts of a Grassmann algebra $\Omega = \Omega^+ + \Omega^-$ :

$$G(g,\Omega) = \exp\left(\Omega^+ \otimes g^+ + \Omega^- \otimes g^-\right) \tag{19}$$

The supergroup $G$ thus depends on the dimensionality of the generating basis $\Omega^1$ of $\Omega$. It can also be generated by the exponentiation of a (reducible) Lie algebra $\tilde{g}$ (the caret in $\hat{\Omega}^\pm$ denotes the basis)

$$\tilde{g} = \hat{\Omega}^+ \otimes g^+ \oplus \hat{\Omega}^- \otimes g^- \tag{20}$$

whose dimensionality $[\tilde{g}]$ depends on that of $\Omega^1$,

$$[\tilde{g}] = \tfrac{1}{2}[g] \times 2^{[\Omega^1]} \tag{21}$$

In our dynamical realization[19] of $G$ = SU(2/1) we work with a specific $\Omega$, the Grassmann algebra of forms over space-time $\Omega(M_{3,1})$. The generators of $\tilde{g}$ will be denoted as $\lambda_u = \mu_u$, $\lambda_i = \mu_i\, dx^\mu$, $\lambda_u^{\mu\nu} = \mu_u\, dx^\mu \wedge dx^\nu, \ldots$ $(\mu_u \leftarrow g^+,\ \mu_i \leftarrow g^-)$ .

The Lie bracket is defined as the exterior product for the forms, times the Lie superbracket $[\mu_M, \mu_N\}$. For instance

$$[\lambda_i^{\ \mu}, \lambda_j^{\ \nu}\} = dx^\mu \wedge dx^\nu \{\mu_i, \mu_j\} = f_i^{\ \mu}{}_j^{\ \nu}{}_{\rho\ \sigma}^{\ a}\ \lambda_a^{\ \rho\sigma} \tag{22}$$

We use square brackets to denote the even subgroup $U(2)_W$ commutation relations, and mixed or curly brackets to denote those relations originating specifically in the superalgebra SU(2/1).

Such a generalized gauge theory involves a generalized system of connections, skew-symmetric contravariant, Bose tensor gauge fields $A_\mu^{\ u}$, $B_{\mu\nu}^{\ \ i}$, $C_{\mu\nu\rho}^{\ \ \ u}$, $E_{\mu\nu\rho\sigma}^{\ \ \ \ i}$ of alternating supergroup gradings, saturating the dimensionality of space-time forms. Under an infinitesimal transformation with parameter $\tilde{\varepsilon}(\varepsilon^u, \varepsilon_\mu^{\ i}, \varepsilon_{\mu\nu}^{\ \ u}\ \varepsilon_{\mu\nu\rho}^{\ \ \ i}, \varepsilon_{\mu\nu\ \rho\sigma}^{\ \ \ \ u})$, the gauge fields vary according to (D denotes the $\lambda_u$ covariant differential with gauge field $A_\mu^u$; d is the exterior differential).

$$\begin{aligned}
\delta A &= -D\varepsilon^{(o)} := -d\varepsilon^{(o)} - [A,\varepsilon^{(o)}] &&=: \tilde{D}\varepsilon^{(o)}\\
\delta B &= -D\varepsilon^{(1)} - [B,\varepsilon^{(o)}] &&=: \tilde{D}\varepsilon^{(1)}\\
\delta C &= -D\varepsilon^{(2)} - \{B,\varepsilon^{(1)}\} - [C,\varepsilon^{(o)}] &&=: \tilde{D}\varepsilon^{(2)}\\
\delta E &= -D\varepsilon^{(3)} - [B,\varepsilon^{(2)}] - [C,\varepsilon^{(1)}] - [E,\varepsilon^{(o)}] &&=: \tilde{D}\varepsilon^{(3)}
\end{aligned} \tag{23}$$

where $A = A_\mu^u\ \mu_a\ dx^\mu$, $B = \frac{1}{2} B_{\mu\nu}^{\ \ i}\ \mu_i\ dx^\mu \wedge dx^\nu$, etc.

These equations define the action of the generalized covariant derivative $\tilde{D}$. The generalized curvature $\tilde{F}$ is similarly defined.

$$\begin{aligned}
&\tilde{F}\ (F^a, G^i, H^a)\\
&F^u = \partial A + \tfrac{1}{2}\,[A,A]^a\\
&G^i = (DB)^i\\
&H^u = (DC)^u + \tfrac{1}{2}\,\{B,B\}^u
\end{aligned} \tag{24}$$

these curvatures transform covariantly,

$$\delta\tilde{F} = [\tilde{\varepsilon},\tilde{F}] \; : \; \delta F = -[F,\varepsilon^{(o)}]$$
$$\delta G = -[G,\varepsilon^{(o)}] - [F,\varepsilon^{(1)}] \qquad (25)$$
$$\delta H = -[H,\varepsilon^{(o)}] - \{G,\varepsilon^{(1)}\} - [F,\varepsilon^{(2)}]$$

and satisfy a Bianchi identity.

The matter multiplets are similarly treated. An irreducible representation $R = R^{+} + R^{-}$ of the superalgebra g will give rise to a representation $\tilde{R}$ of $\tilde{g}$. Denote by $\tilde{\phi}$ a system of 0,1,2... forms taking their values alternatively in $R^{+}$ and $R^{-}$, $\tilde{\phi}$ $(\phi^{U}, \psi_{\mu}{}^{I}dx^{\mu}, \Xi_{\mu\nu}{}^{U}dx^{\mu} \wedge dx^{\nu},..)$. The representation is defined by the transformation rules,

$$\delta\,\phi = [\varepsilon^{(o)},\,\phi]$$
$$\delta\,\psi = [\varepsilon^{(o)},\,\phi] + \{\varepsilon^{(1)},\,\phi\} \qquad (26)$$
$$\delta\,\Xi = [\varepsilon^{(o)},\,\Xi] + \{\varepsilon^{(1)},\,\psi\} + [\varepsilon^{(2)},\,\phi]$$

with the Jacobi identity automatically satisfied.

The free field Lagrangian has the same physical degrees of freedom as the Weinberg-Salam model. Given a skew p-tensor $\phi_{\mu\nu...}$ and its generalized curl

$$L_{p} = -\frac{1}{2p!}(\delta_{[\mu}\,\phi_{\nu\rho..]})^{2} \qquad (27)$$

The number of physical degrees of freedom n for the gauge fields[23] is precisely given by adding up the number $\binom{N}{k}$ of components of an antisymmetric k-indices tensor in N dimensions, together with the number of dy-contracted components of its complexified[18)] geometrical vertical complements (in the direction of the fibers $y^{M},y^{\bar{M}}$ in the bundle manifold, ghosts counting negatively). Denoting the fiber-complexified forms by a caret,

$$\hat{A}^u = A^u_\mu \, dx^r + A^u_M \, dy^M + A^u_{\bar{N}} \, dy^{\bar{N}} = A^u_\mu \, dx^\mu + \alpha^u + \bar{\alpha}^u$$

$$\begin{aligned}\hat{B}^i &= \tfrac{1}{2} B^{\ i}_{\mu\nu} \, dx^\mu \wedge dx^\nu + B^i_{\mu M} \, dx^\mu \wedge dy^M + B^i_{\mu\bar{N}} \, dx^\mu \wedge dy^{\bar{N}} \\ &+ \tfrac{1}{2} B^i_{MN} \, dy^M \wedge dy^N + B^i_{M\bar{N}} \, dy^M \wedge dy^{\bar{N}} + \tfrac{1}{2} B^i_{\overline{MN}} \, dy^{\bar{M}} \wedge dy^{\bar{N}} \\ &= \tfrac{1}{2} B^{\ i}_{\mu\nu} \, dx^\mu \wedge dx^\nu + \beta^i_\mu \, dx^\mu + \bar{\beta}^i_\mu dx^\mu + b^i + h^i + \bar{b}^i, \text{ etc...}\end{aligned} \tag{28}$$

Latin letters denote Bose fields; Greek, Fermi ghost fields. We have (per internal index)

$$\begin{aligned} n(A) &= \binom{4}{1} - (2 \times 1) = 2 \\ n(B) &= \binom{4}{2} - (2 \times 4) + 3 = 1 \\ n(C) &= \binom{4}{3} - (2 \times 6) + (3 \times 4) - 4 = o \quad \text{etc...}\end{aligned} \tag{29}$$

We note that $\hat{B}^i$ contains a scalar real $h^i$ multiplet, required in the SU(2/1) irreps $\underset{\sim}{8}(\alpha^a, h^i)$ or $\underset{\sim}{\bar{8}}(\bar{\alpha}^a, h^i)$, in the Curci-Ferrari type symmetric-complexified algebra of ghosts. The higher forms C and E do not contribute to the physical spectrum.

We have shown[19,20] how in this formalism the matter fields acquire ghosts with inverted chiralities.

## 6. THE INTERACTING LAGRANGIAN

The Lie algebra is reducible and its Killing metric is non zero only in the $A^u_\mu$ sector, so only the Yang-Mills vector Lagrangian comes out as a natural invariant. On the other hand, if we consider as a Lagrangian for the $B^i_{\mu\nu}$ the term

$$L_B = -\tfrac{1}{2} \, (G^i_{\mu\nu\rho})^2 \tag{30}$$

using the $U(2)_w$ $\delta_{ij}$ metric, the $B^i_{\mu\nu}$ equation of the motion $D_\mu G_{\mu\nu\rho} = o$ enforces the constraint

$$D_\mu D_\nu G_{\mu\nu\rho} = [F_{\mu\nu}, G_{\mu\nu\rho}] = o \tag{31}$$

The natural invariance of the Lagrangian under $U(2)_w$ covariant BRS variations $SB_{\mu\nu} = D_{[\mu}\beta_{\nu]}$ is also lost if the constraint is not satisfied:

$$sL_B = G_{\mu\nu\rho} D_\mu D_\nu \beta_\rho = [G_{\mu\nu\rho}, F_{\mu\nu}] \beta_\rho \qquad (32)$$

where $\beta_\nu$ is the vector ghost in (28) and

$$S = s + [\alpha, ] \qquad (33)$$

We introduce a Lagrange multiplier $K_\mu$ whose equation of the motion enforces the differential constraints, and consider the Lagrangian

$$L'_B = - {}^1/_{12} (\varepsilon^{\mu\nu\rho\tau}(D_\mu(B_{\nu\rho} + D_\nu K_\rho)))^2 \qquad (34)$$

The equations of motion are

$$D_\mu(D_{[\mu}(B_{\nu\rho]} + D_\nu K_{\rho]})) = o$$
$$[F_{\mu\nu}, D_{[\mu} B'_{\nu\rho]}] = o \qquad (35)$$

and the system is now closed. At the same time, $L'_B$ is invariant under the nilpotent BRS algebra involving fields and ghosts from (28), with $\kappa$ as the ghost of $K_\mu$, (note the role of the $\beta_\mu$ of rep.$(o,1^+)$).

$$S\alpha = \tfrac{1}{2}[\alpha,\alpha]$$
$$SA_\mu = \partial_\mu \alpha$$
$$Sb = o$$
$$S\kappa := -b \qquad (36)$$
$$S\beta_\mu = D_\mu b$$
$$S K_\mu = -(\beta_\mu + D_\mu \kappa) := -\beta'_\mu$$
$$S B_{\mu\nu} = D_{[\mu}\beta_{\nu]} + [F_{\mu\nu},\kappa] = D_{[\mu}\beta'_{\nu]}$$

Our method admits a direct generalisation to the $C^a_{\mu\nu\rho}$ field; $E^i_{\mu\nu\rho\sigma}$ has no curl in 4 dimensions. The complete classical $\tilde{g}$ gauge Lagrangian can be written, using an auxiliary 2-form $\tilde{L}(L_{\mu\nu}, \lambda_\mu, l)$

$$L = -\tfrac{1}{4}(F_{\mu\nu}(A))^2 - {}^1/_{12}(D_{[\mu} B'_{\nu\rho]})^2$$
$$- {}^1/_{48}(D_{[\mu} C'_{\nu\rho\sigma]} + \tfrac{1}{2}\{B'_{\mu\nu}, B'_{\rho\sigma}\})^2 \qquad (37)$$

$$B'_{\mu\nu} := B_{\mu\nu} + D_{[\mu}K_{\nu]} \ , \quad C'_{\mu\nu\rho} := C_{\mu\nu\rho} + D_{[\mu}L_{\nu\rho]}$$

The squares are computed using the $\delta_{ab}$, $\delta_{ij}$ metric, implying

$\sin^2\theta_W = .25$. The dynamics correspond to the gauging of an even Lie algebra $\tilde{g}$, and the Killing metric of SU(2/1) plays no role here.

This Lagrangian is invariant under the nilpotent BRS algebra given above for the $C_{\mu\nu\rho}$ sector:

$$\begin{aligned}
S\gamma &= \tfrac{1}{2}\{b,b\} \quad , \quad S\ell = -\gamma - \tfrac{1}{2}\{\kappa,b\} := -\gamma' \\
Sc'_\mu &= D_\mu \gamma' + \{\beta',b\} \\
S\lambda_\mu &= -c_\mu - D_\mu \ell - \{K_\mu,b\} := -c'_\mu \\
S\Gamma_{\mu\nu} &= D_{[\mu} c'_{\nu]} + \tfrac{1}{2}\{\beta'_{[\mu},\beta'_{\nu]}\} + \{B'_{\mu\nu},b\} \\
SL_{\mu\nu} &= -\Gamma_{\mu\nu} - D_{[\mu}\lambda_{\nu]} - \{B'_{\mu\nu},\kappa\} - \tfrac{1}{2}\{\beta'_{[\mu},K_{\nu]}\} := -\Gamma'_{\mu\nu} \\
SC_{\mu\nu\rho} &= D_{[\mu}\Gamma'_{\nu\rho]} + \{B'_{[\mu\nu},\beta'_{\rho]}\}
\end{aligned} \tag{38}$$

At the linearized level, we recover the Weinberg-Salam spectrum.

Clearly, we still have to investigate the manner in which the spontaneous symmetry breakdown should be implementated. However, the fit between SU(2/1) and the selection of multiplets and parameters for the electroweak theory is remarkable.

REFERENCES

1. P. Reutens et al, in The Santa Fe Meeting, edited by T. Goldman and M. M. Nieto, World Scientific Pub., Philadelphia-Singapore, pp. 270-273 (1985).

2. Y. Ne'eman, Phys. Lett. B81, 190 (1979).
D. B. Fairlie, Phys. Lett. B82, 97 (1979).

3. Y. Ne'eman, to be pub. in Proc. 14th International Colloquium on Group-Theoretical Methods in Physics (Seoul 1985), World Scientific Pub., Singapore.

4. H. Georgi and S. L. Glashow, Phys. Rev. Lett. 32, 438 (1974).

5. H. Georgi, H. R. Quinn and S. Weinberg, Phys. Rev. Lett. 33,451 (1974).

6. S. P. Rosen, in ref.[1], pp. 171-191.

7. Y. Ne'eman, in Cosmology and Gravitation, Spin, Torsion, Rotation and Supergravity (Erice 1979), P. G. Bergmann and V. de Sabbata eds., Plenum Press (N.Y. and London, 1980) Nato Advanced Institutes, Series B: Physics 58, pp. 177-226.

8. M. Gell-Mann and Y. Ne'eman, The Eightfold Way, W. A. Benjamin, Pub., N.Y. (1964).

9. V. Rittenberg, in Group Theoretical Methods in Physics, Prov. VI. International Conference (Tübingen, 1977), P. Kramer and A. Rieckers eds., Springer-Verlag Lecture Notes in Physcs 79, Berlin-Heidelberg-N.Y., pp. 3-21 (1978).

10. Y. Ne'eman, in Differential Geometrical Methods in Mathematical Physics, P. L. Garcia et al editors, Springer Verlag Lecture Notes in Mathematics 836, Berlin-Heidelberg-N.Y., pp. 318-348 (1980).

11. M. Scheunert, W. Nahm and V. Rittenberg, Jour. Math. Phys. 18,155 (1977).

12. Y. Ne'eman and S. Sternberg, Proc. Nat. Acad. Sc. USA, 77, pp.3127-3131 (1980).

13. Y. Ne'eman and J. Thierry-Mieg, Phys. Lett. B108, 399-402 (1982).

14. Y. Ne'eman, S. Sternberg and J. Thierry-Mieg, Proc. Rome (GUD 1981) Workshop, G. Cignetti et al editors, SDLN Frascati Pub. pp.89-92 (1982).

15. C. Becchi, A. Rouet and R. Stora, Comm. Math. Phys. 42, 126 (1975).

16. G. Curci and R. Ferrari, Nuovo Cimento 30A, 155-168 (1975).

17. Y. Ne'eman, in Unification of the Fundamental Interactions, S. Ferrara et al editors, Plenum Press (New York, N.Y.) pp. 89-94 (1980).

18. J. Thierry-Mieg and Y. Ne'eman, Nuovo Cimento 71A, 104-118 (1982).

19. J. Thierry-Mieg and Y. Ne'eman, Proc. Nat. Acad. Sci., USA, 79, pp. 7068-7072 (1982).

20. J. Thierry-Mieg and Y. Ne'eman, in Differential Geometric Methods in Mathematical Physics, S. Sternberg ed., Riedel Pub., Amsterdam, pp. 101-114 (1984).

# FERMION FRACTIONIZATION AND SUPERSYMMETRY

Shoichi Midorikawa
Institute for Nuclear Study,
University of Tokyo
Tanashi, Tokyo 188, Japan

Abstract

Fractional fermion numbers induced on solitons are investigated by using the canonical quantization procedure. Especially, charge conjugation operator is constructed in terms of creation and annihilation operators. We also discuss a "hidden" supersymmetry in the 3+1 dimensional monopole-fermion system.

## 1. INTRODUCTION

Since the discovery by Jackiw and Rebbi,[1] solitons with fractional fermion numbers have been investigated extensively.[2] If the system has charge conjugation symmetry, the fermion number of this system must be either integer or a half-integer. Jackiw and Rebbi have realized the latter example in the 1+1 and 3+1 dimensional models by showing that the expectation value of the fermion number induced on a soliton is a half-integer.

The next question that arises is whether the fractional charge is a definite observable, i.e., whether a soliton is an eigenstate of the charge operator. This question has been answered affirmatively.[3] Consider a soliton — anti-soliton system which has integer eigenvalues for the charge operator. In the limit of infinite soliton anti-soliton separation, it is possible for the soliton to become an eigenstate of the charge operator by making a Bogoliubov transformation. In this limit, the localized charge operator has fractional eigenvalues.

It remains to be seen how to construct the charge conjugation operator in the monopole fermion system. In this note, we shall discuss this problem. As has been discussed above, the charge conjugation invariance makes an important role for a system to have a half-integer charge. In sec. 2, we shall consider how to construct a charge conjugation operator in terms of creation and annihilation operators.

We shall also discuss "hidden" supersymmetry of the Dirac Hamiltonian.[4)-7)] We shall show in sec. 3 that a fermion zero mode around a 't Hooft-Polyakov monopole[8)] arises as a consequence of the "hidden" supersymmetric quantum mechanics although the original supersymmetric quantum mechanics[9)-11)] is defined in one-dimensional space. Thus this phenomenon can be interpreted as the manifestation of the supersymmetric quantum mechanics in three dimensional space proposed by Takeda and Ui.[12)]

## 2. CHARGE CONJUGATION

The 1+1 dimensional model considered here involves a scalar field $\phi$ and a spinor $\psi$, with a Lagrangian density of the form

$$\mathscr{L} = \frac{1}{2}(\partial_\mu\phi)(\partial^\mu\phi) - V(\phi) + i\bar{\psi}[\not{\partial}+ig\phi]\psi \ . \tag{2.1}$$

The $\gamma$-matrices in our convension are $\gamma^0 = \sigma_2$ and $\gamma^1 = i\sigma_1$. In this representation, charge conjugation is simply to take complex conjugation, i.e.,

$$\psi_c = C\bar{\psi}^T = \psi^* , \tag{2.2}$$

where $C = \gamma^0$ and Majorana spinors satisfy $\psi = \psi^*$. We further suppose that in the absence of fermions, the static field equation for $\phi$ possesses a soliton solution $\phi_c(x)$ which satisfies $\phi_c(\pm\infty) = \pm\mu$. The Dirac equation for the fermion field operator $\psi$ in this background field is given by

$$[i\partial\!\!\!/ - g\phi_c]\psi = 0 . \tag{2.3}$$

This equation has only one zero-energy bound state of

$$\psi_0(x) = N \exp[-g\int_0^x \phi_c(y)dy]\begin{pmatrix}1\\1\end{pmatrix} . \tag{2.4}$$

We assume that there is no other bound states and express the continuum states associated with the positive energy states, $u(p)$, and with the negative energy states, $v(p) = e^{-i\varphi(p)}u^*(p)$.

The quantum state of the fermion field is obtained by the following eigenmode expansion:

$$\psi = b_B^\dagger\psi_0 + \int\frac{dp}{2\pi}\sqrt{\frac{m}{E_p}}\,[a(p)u(p)e^{-ip\cdot x}+b^\dagger(p)v(p)e^{ip\cdot x}] . \tag{2.5}$$

Here $a(p)$ and $b(p)$ are annihilation and creation operators while $b_B^\dagger$ is associated with the zero-energy solution and there is no requirement that $b_B^\dagger$ and $b_B$ are creation and annihilation operators. The anticommutation relations of $\psi$ requires

$$\{a(p),a^\dagger(p)\} = \{b(p),b^\dagger(p)\} = \{b_B,b_B^\dagger\} = 1 \tag{2.6}$$

with all other anticommutators vanishing.

Now we seek a unitary operator $\mathcal{C}$ of charge conjugation. From eq. (2.2), this operator generates the following transformation:

$$\psi^* = \mathcal{C}\psi\mathcal{C}^{-1} \tag{2.7}$$

Recall that positive and negative energy states are related by

$$u^*(p) = v(p)e^{i\varphi(p)} \quad \text{and} \quad v^*(p) = u(p)e^{i\varphi(p)} \tag{2.8a,b}$$

In general, the zero-energy state satisfies

$$\psi_0^* = e^{i\varphi_B}\psi_0 \tag{2.8c}$$

by an appropriate choice of N in eq. (2.4). Carrying out the momentum expansion (2.5) of eq. (2.7), we obtain

$$\mathcal{C}b_B^\dagger\mathcal{C}^{-1} = b_B e^{i\varphi_B} , \tag{2.9a}$$

$$\mathcal{C}a(p)\mathcal{C}^{-1} = b(p)e^{i\varphi(p)} , \qquad \mathcal{C}b^\dagger(p)\mathcal{C}^{-1} = a^\dagger(p)e^{i\varphi(p)} . \tag{2.9b,c}$$

It is convenient to decompose $\mathcal{C}$ into a product of two unitary transformations as

$$\mathcal{C} = \mathcal{C}_2\mathcal{C}_1 \tag{2.10}$$

and to choose $\mathcal{C}_1$ to remove the phase factor $\varphi$ as

$$\mathcal{C}_1 b_B^\dagger \mathcal{C}_1^{-1} = e^{i\varphi_B} b_B^\dagger \tag{2.11a}$$

$$\mathcal{C}_1 a(p) \mathcal{C}_1^{-1} = -e^{i\varphi(p)} a(p) \ , \qquad \mathcal{C}_1 b^\dagger(p) \mathcal{C}_1^{-1} = -e^{i\varphi(p)} b^\dagger(p) \ . \tag{2.11b,c}$$

By an explicit construction, we find

$$\mathcal{C}_1 = \exp\, i[\varphi_B b_B^\dagger b_B + \int dp(\pi - \varphi(p))(a^\dagger(p)a(p) - b^\dagger(p)b(p))]\} \tag{2.12}$$

Also, we find

$$\mathcal{C}_2 = (b_B^\dagger + b_B) \exp\, \{\frac{i\pi}{2} \int dp [a^\dagger(p) - b^\dagger(p)][a(p) - b(p)]\} \tag{2.13}$$

It should be noted that we can define a charge conjugation operator in a usual way in a soliton — anti-soliton system by taking the infinite separation of the solitons. However, the operator constructed in this way is different from the above obtained $\mathcal{C}$. The difference comes from the number of ground states. In a soliton system the ground states are doubly degenerate while in the limiting procedure a full use of the quadratically degenerate ground states must be made.

## 3. "HIDDEN" SUPERSYMMETRY

It is easy to see how the previous example contains supersymmetry as a "hidden" symmetry. Applying $i\not{\partial} + g\phi_c$ into eq. (2.3), we obtain[6]

$$-\frac{\partial^2}{\partial t^2}\psi = [-\frac{\partial^2}{\partial x^2} + g^2\phi_c^2 - \sigma_1 g \frac{d\phi_c}{dx}]\psi \tag{3.1}$$

The right hand side of eq. (3.1) has the form in the supersymmetric quantum mechanics. If one wants to diagonalize the spin interaction term, one should use $e^{i\frac{\pi}{4}\sigma_2}\psi$ instead of $\psi$.

Next, we shall search for a "hidden" supersymmetry in a 3+1 dimensional theory, i.e., a theory of the SU(2) monopoles.[8] This may serve as the condition in which fermions have zero energy mode. As is known, the isovector fermions coupled with the Yang-Mills field of the monopole constitute the N=1 vector superfield[4] in the Prasad-Sommerfield limit.[13] We shall consider as another example of the isospinor fermions coupled with the monopole.

The Dirac equation in the external potential is

$$[i\not{D}_{nm} - \frac{1}{2} G\tau^a_{nm}\Phi_a]\Psi_m = 0 , \qquad (3.2)$$

where $\not{D}$ is the covariant derivative with the gauge coupling constant g and G is the Yukawa coupling constant. As the monopole solution, we will choose

$$A^0_a = 0 , \quad \Phi_a = \frac{r_a}{r}\phi(r) , \quad A^i_a = \epsilon^{aij}\frac{r_j}{gr}A(r) \qquad (3.3)$$

where the backreaction of the fermions are ignored. Let us use the following representation of the Dirac matrices:

$$\alpha = \begin{bmatrix} 0 & \sigma \\ \sigma & 0 \end{bmatrix} \qquad \beta = -i\begin{bmatrix} 0 & I \\ -I & 0 \end{bmatrix} \qquad (3.4)$$

Let us also decompose $\Psi_n$ into upper and lower components as

$$\Psi_n = \begin{bmatrix} \chi_n^+ \\ \chi_n^- \end{bmatrix}. \tag{3.5}$$

Further, let us represent $\chi^\pm$ with 2×2 matrices $\mathcal{M}^\pm$ as

$$\chi_{in}^\pm = \mathcal{M}_{in}^\pm (\tau_2)_{mn} . \tag{3.6}$$

Define the operator $\vec{J} = \vec{j} + \vec{I}$, where $\vec{j}$ is the total angular momentum and $\vec{I}$ is the isospin. In the $\vec{J} = 0$ case, we can expand $\mathcal{M}^\pm$ in terms of four scalar functions as

$$\mathcal{M}_{im}^\pm(\vec{r}) = g^\pm(r)\delta_{im} + f^\pm(r) \frac{r_a}{r} \sigma_{im}^a . \tag{3.7}$$

Substituting (3.3)-(3.7) into (3.2), we obtain

$$\left[ \frac{d}{dr} + \frac{1}{r} - (A + \frac{1}{r} \mp \frac{1}{2} G\phi)\tau_3 \right] \begin{pmatrix} g^\pm \\ f^\mp \end{pmatrix} = iE\tau_1 \begin{pmatrix} g^\pm \\ f^\mp \end{pmatrix} \tag{3.8}$$

where E is the energy of the fermion. It is convenient to define $u_\pm$ by

$$u_\pm = r \begin{pmatrix} g^\pm \\ f^\mp \end{pmatrix} . \tag{3.9}$$

Then, using the operators

$$Q_\pm = -i\frac{d}{dr} \tau_1 + (A + \frac{1}{r} \mp \frac{1}{2} G\phi)\tau_2 , \tag{3.10}$$

we can we write eq. (3.8) as follows:

$$Q_{\pm}u_{\pm} = Eu_{\pm} \ . \tag{3.11}$$

Applying $Q_{\pm}$ into eq. (3.10), we obtain

$$E^2 u_{\pm} = [- \frac{d^2}{dr^2} + W_{\pm}^2 + \tau_3 \frac{dW_{\pm}}{dr} ] u_{\pm} \ , \tag{3.12}$$

where

$$W_{\pm}(r) = A(r) + \frac{1}{r} \mp \frac{1}{2} G\phi(r) \ . \tag{3.13}$$

Thus, $Q_{\pm}$ are recognized as the supercharge and the operators in the right hand side of eq. (3.12) are regarded as the Hamiltonian in the supersymmetric quantum mechanics.

Whether supersymmetry is broken spontaneously[9)] is determined by the number of zeros of $W_{\pm}$ in eq. (3.13). If $W_{\pm}$ has an odd number of zeros, there exists only one supersymmetric state. This means that there is a zero-energy fermion bound state around the monopole. As an illustration, let us consider the Prasad-Sommerfield limit where A and $\phi$ is given by

$$A = \frac{1}{r}[\frac{Mr}{\sinh(Mr)} - 1] \quad \text{and} \quad \phi = [M\coth(Mr) - \frac{1}{r}] \ . \tag{3.14}$$

In this case, there is a solution of the equation $W_{+}(r) = 0$, while no solution of $W_{-}(r) = 0$.

We have shown that the supersymmetry is dynamically generated in a wide range of monopole-fermion systems, though we have not found the right way to use this supersymmetry.

ACKNOWLEDGMENT

The author would like to thank Prof. H. Terazawa for reading the manuscript.

REFERENCES

1) R. Jackiw and C. Rebbi, Phys. Rev. D13 (1976) 3398.
2) For a review, A. J. Niemi and G. W. Semenoff "Fermion Number Fractionization in Quantum Field Theory", IAS preprint (1984).
3) S. Kivelson and J. R. Schrieffer, Phys. Rev. B25 (1982) 6447; R. Jackiw, A. K. Kerman, I. Klevabanov, and G. Semenoff, Nucl. Phys. B225 [FS9] (1983) 233.
4) P. Rossi, Phys. Lett. 71B (1977) 145;
5) G. Semenoff, H. Matsumoto, and H. Umezawa, Phys. Rev. D25 (1982) 1045; Phys. Lett. 113B (1982) 371.
6) M. Stone, Ann. of Phys. 155 (1984) 56.
7) S. Midorikawa, Phys. Lett. 138B (1984) 111.
8) G. 't Hooft, Nucl. Phys. B79 (1974) 276; A. M. Polyakov, ZhETF Pis. Red. 20 (1974) 430 [JETP Lett. 20 (1974) 194].
9) E. Witten, Nucl. Phys. B185 (1981) 513.
10) P. Salomonson and J. W. van Holten, Nucl. Phys. B196 (1982) 509.
11) F. Cooper and B. Freedman, Ann. of Phys. 146 (1983) 262.
12) G. Takeda and H. Ui, Prog. Theor. Phys. 73 (1985) 1061.
13) M. K. Prasad and C. M. Sommerfield, Phys. Rev. Lett. 35 (1975) 760.

# EFFECT OF EXTRA DIMENSIONS IN SELF-GRAVITATING SPHERICAL BODIES

M. Yoshimura

National Laboratory for High Energy Physics (KEK),

Oho-machi, Tsukuba-gun, Ibaraki-ken 305 Japan

ABSTRACT

We initiate study of three dimensional spherically symmetric metrics in which the size of an extra compact manifold is allowed to vary with distance. After establishing a fundamental set of equations, we find a few nontrivial exact solutions. In the first example uniqueness of the black hole solution is violated due to a Ricci-flatness of the internal manifold. In the second example a dense uniform star can be bound, exceeding the well known upper limit of the red shift, $z = 2$. Although realistic solutions are yet to be found, these solutions serve to illustrate peculiar features of higher dimensional gravity viewed from four spacetime dimension.

The fundamental equation of spacetime structure has been presumed to be the Einstein equation until very recently. The ordinary form of general relativity is, however, probably not consistent with quantum mechanics, as exemplified by the Hawking radiation. More recently, search for anomaly-free gravitational theories led to superstring theories[1)] that are well defined only in ten dimensions and that can incorporate two particular gauge groups, SO(32) or $E_8 \times E_8$. To be a realistic theory, six dimensions out of the total ten have to be compactified to a size of order the Planck length. Thus, the theory resembles the old Kaluza-Klein idea of general relativity in more than four dimensions although the theory contains many more fields. Remarkably, existence of extra gauge fields in these superstring theories also solves the long standing problem of Kaluza-Klein theories, the problem of massless chiral fermion[2)].

Developments of the superstring theory are as yet in a primitive stage. In particular, we do not know the underlying geometrical picture that gives rise to gauge invariances of gravity and Yang-Mills interactions. Nonetheless, we may already have a glimpse of how the Einstein equation should be modified at short distances. The basic reason for this is that we have presumably correct, computational rules[3)] of evaluating tree and loop amplitudes of string theories in the particular light cone gauge, at least on Ricci-flat background manifolds. An effective ten dimensional field theory thus emerges by systematically expanding the string theory in powers of the string size, $(\alpha')^{1/2}$. (This simple expansion is justified only for the closed string model, the heterotic string. Theories with the open string are much more complicated.)

In view of these recent developments, exploration[4)] of physical implications of the extra dimension, that has been undertaken for some time prior to the advent of superstring theories, becomes even more urgent. In this paper we shall investigate three dimensional spherically symmetric static solutions of higher dimensional gravity implied by the superstring theory. These solutions are hoped to be relevant in formation of black holes. Cosmological implications in a time dependent Robertson-Walker type of metric, with a kind of compact manifold favored by superstring phenomenology, have been studied in another publication[5].

One of the most interesting aspects in superstring theories is that the extra six dimensional manifold is likely to be Ricci-flat; the Ricci-flat Calabi-Yau space[6] is favored by the gauge hierarchy problem and in the filed theory limit of the heterotic string the Calabi-Yau compactification is consistent with field equations when a computed $R_{ABCD} R^{ABCD}$ term is included[7]. In the first part of this paper we shall establish a fundamental set of equations when the size of the extra space is allowed to vary with the spherical symmetry. We shall then discuss a peculiar feature of gravity when a Ricci-flat compact internal manifold is present. One could most dramatically express this peculiarity by saying that the uniqueness of the black hole solution in the spherically symmetric case is violated. We show this by giving explicit vacuum solutions in which the Birkoff theorem is not applied. This peculiar feature is caused by a long range force associated with the Ricci-flatness of the manifold. In a realistic compactification supersymmetry breaking at a lower mass scale m ($\gtrsim$ 100 GeV) will occur[8], and this would provide a cutoff of the long range force at a length scale of order $m^{-1}$. Thus, effect of the non-uniqueness of the black hole will be very small at the scale of astronomical bodies. But the long range force might significantly modify formation of primordial black holes at the very early universe. In the second part of this paper we shall discuss spherically symmetric solutions in presence of matter. The bound on the surface gravity, hence on the maximum red shift factor, in general relativity can be violated in this case.

To appreciate the role of extra dimensions, it is most convenient to view the higher dimensional gravity as a field theory in four dimensions. We often consider the following class of metric in problems of this kind as well as in cosmology,

$$ds^2 = g_{ij}(x)\, dx^i dx^j - b^2(x)\, \tilde{g}_{\alpha\beta}(y)\, dy^\alpha dy^\beta . \tag{1}$$

We use the signature convention of (+---...). The coordinate y is used to label different points in a compact manifold specified by the metric $\tilde{g}_{\alpha\beta}$, while the coordinate x labels points in the usual four dimensional specetime. The lower case Latin and Greek indices refer to the four and the extra d-dimensional spaces, respectively. The scale factor b(x) can be regarded as a position dependent size of the extra manifold. For the metric (1) nonvanishing components of the connection are $\Gamma^i_{jk}$ (x), $\Gamma^\alpha_{\beta\gamma}$ (y) and

$$\Gamma^i_{\alpha\beta} = b\, b_,{}^i \tilde{g}_{\alpha\beta}, \quad \Gamma^\alpha_{\beta i} = b_{,i}\, b^{-1} \delta^\alpha{}_\beta . \tag{2}$$

Nonvanishing Riemann curvature tensors are then

$$R^i{}_{jk\ell}(x),\ R^i{}_{\alpha j\beta} = -b\, b_;{}^i{}_j\, \tilde{g}_{\alpha\beta} \ ,$$

$$R^\alpha{}_{\beta\gamma\delta} = \tilde{R}^\alpha{}_{\beta\gamma\delta}(y) + b_{,i}\, b_,{}^i (\tilde{g}_{\beta\gamma} \delta^\alpha{}_\delta - \tilde{g}_{\beta\delta} \delta^\alpha{}_\gamma). \tag{3}$$

Our convention of the curvature tensor is such that $R^A{}_{BCD} = \Gamma^A_{BC,D} + \ldots$, $R_{AB} = R^C{}_{ABC}$. The covariant derivative ; i is defined with respect to the metric $g_{ij}$ (x), and the x-independent curvature $\tilde{R}^\alpha{}_{\beta\gamma\delta}$ (y) is computed from the metric of the extra space, $\tilde{g}_{\alpha\beta}$ (y).

The Einstein-Hilbert action with the cosmological term included is then reduced to a four dimensional action after integrating out internal degrees of freedom,

$$- (2\kappa^2)^{-1} \int dx dy \sqrt{g}\ (R + 2\Lambda)$$

$$= -(2\kappa^2)^{-1} v_d \int dx\ \sqrt{g_4}\ b^d(x)\ [R^{(4)}(x) - d(d-1)\ b^{-2} b_{,i} b^{\ i}_{,}$$

$$-2d\ b^{-1} b_{;i}^{\ \ i} - c\ b^{-2} + 2\Lambda]\ , \tag{4}$$

$$c \equiv v_d^{-1} \int dy\ (\tilde{g})^{1/2}\ \tilde{R}\ (y), \tag{5}$$

$$v_d \equiv \int dy\ (\tilde{g})^{1/2}, \qquad g_4 \equiv \det\ [-g_{ij}(x)]\ . \tag{6}$$

The curvature $\tilde{R}$ (y) is a scalar curvature computed from $\tilde{g}_{\alpha\beta}$ (y) alone, and the average curvature constant c is usually nonnegative for compact manifolds. d is a dimension of the compact manifold. After a Weyl rescaling, $g_{ij}(x) \rightarrow b^d(x)\ g_{ij}(x)$, the action becomes

$$(2\kappa^2)^{-1} v_d \int dx\ \sqrt{g_4}\, [-R^{(4)}(x) + \tfrac{1}{2} d(d+2)\ b^{-2} b_{,i} b^{\ i}_{,}$$

$$+ cb^{-d-2} - 2\Lambda\ b^{-d} + (\text{total derivative})]\ . \tag{7}$$

It is thus easy to see that the size of the extra dimension b(x) at each spacetime point plays the role of a scalar field in four dimensions. More precisely, the correctly normalized scalar field is given by

$$\phi\ (x) = [\tfrac{1}{2}\ d(d+2) v_d]^{1/2} \kappa^{-1} \ell n\ b(x)\ . \tag{8}$$

In this four dimensional picture the curvature term of the extra compact manifold and the $\Lambda$-term act as a potential of the scalar field $\phi$. Ricci-flat manifolds such as a torus and Calabi-Yau spaces[6)] are special in the sense that they give rise to a massless scalar field when the cosmological constant is absent.

Consider now a general form of spherically symmetric metric in which the extra size b(r) is also a function of three dimensional distance r from a center,

$$ds^2 = e^\sigma\ dt^2 - e^\omega\ dr^2 - r^2\ (d\theta^2 + \sin^2\theta\ d\phi^2) - b^2 d\Omega, \tag{9}$$

with $d\Omega = \tilde{g}_{\alpha\beta}(y)\, dy^\alpha dy^\beta$. $\sigma$, $\omega$ and b are functions of r. Nonvanishing components of the connection and the Ricci tensor are then computed as ($' = d/dr$)

$$\Gamma^r_{oo} = \frac{1}{2}\sigma'\, e^{\sigma-\omega},\ \Gamma^o_{or} = \frac{1}{2}\sigma',\ \Gamma^r_{rr} = \frac{1}{2}\omega',\ \Gamma^r_{\theta\theta} = -re^{-\omega},$$

$$\Gamma^r_{\phi\phi} = -r\sin^2\theta\, e^{-\omega},\ \Gamma^\theta_{\theta r} = \Gamma^\phi_{\phi r} = \frac{1}{r},\ \Gamma^\phi_{\phi\theta} = ct\theta,\ \Gamma^\theta_{\phi\phi} = -\sin\theta\cos\theta,$$

$$\Gamma^r_{\alpha\beta} = -b\, b'\, e^{-\omega}\, \tilde{g}_{\alpha\beta},\ \Gamma^\alpha_{\beta r} = b^{-1}\, b'\, \delta^\alpha{}_\beta, \tag{10}$$

$$R_{oo} = e^{\sigma-\omega}\left(\frac{1}{2}\sigma'' + \frac{\sigma'}{r} + \frac{1}{4}\sigma'^2 - \frac{1}{4}\sigma'\omega' + \frac{d}{2}b^{-1}\sigma' b'\right), \tag{11}$$

$$R_{rr} = -\frac{1}{2}\sigma'' - \frac{1}{4}\sigma'^2 + \frac{1}{4}\sigma'\omega' + \frac{\omega'}{r} - d\, b^{-1}\, b'' + \frac{d}{2}b^{-1}\omega' b', \tag{12}$$

$$R_{\theta\theta} = \sin^{-2}\theta\, R_{\phi\phi} = \frac{1}{2}r^2 e^{-\omega}\left(-\frac{\sigma'}{r} + \frac{\omega'}{r} - 2db^{-1}\frac{b'}{r} - \frac{2}{r^2}\right) + 1, \tag{13}$$

$$R_{\alpha\beta} = \tilde{R}_{\alpha\beta}(y) - \tilde{g}_{\alpha\beta}e^{-\omega}\{b\, b'' + (d-1)\, b'^2 - \frac{1}{2}bb'\omega'$$
$$+ \frac{1}{2}bb'\sigma' + \frac{2}{r}bb'\}. \tag{14}$$

Besides size dependent terms containing b(r), the scalar curvature given here, of cource, agrees with well known expressions[9]. $\tilde{R}_{\alpha\beta}(y) = \tilde{R}^\gamma{}_{\alpha\beta\gamma}(y)$ is the Ricci tensor of the extra manifold with b = 1.

Let us assume a general form of anisotropic energy-momentum tensor,

$$(T_{AB}) = (\rho\, g_{oo},\ -p_1 g_{ij},\ p_2 b^2\, \tilde{g}_{\alpha\beta}), \tag{15}$$

$g_{ij}$ = spatial part of the four dimensional metric, and the energy density $\rho$ and the pressures $p_i$ may contain coherent contribution of field condensates and the $\Lambda$-term. The Einstein equation, $R_{AB} = \kappa^2\{T_{AB} - (d+2)^{-1} g_{AB}\, T^C{}_C\}$, reads as

$$R_{oo} = \kappa^2 e^\sigma (d+2)^{-1}\{(d+1)\rho + 3p_1 + dp_2\}, \tag{16}$$

$$R_{rr} = \kappa^2 e^\omega (d+2)^{-1}\{\rho + (d-1)p_1 - dp_2\}, \tag{17}$$

$$R_{\theta\theta} = \sin^{-2}\theta\, R_{\phi\phi} = \kappa^2 r^2 (d+2)^{-1}\{\rho + (d-1)p_1 - dp_2\}, \tag{18}$$

$$R_{\alpha\beta} = \kappa^2 b^2\, \tilde{g}_{\alpha\beta}(d+2)^{-1}(\rho - 3p_1 + 2p_2). \tag{19}$$

Clearly, the ansatz (15) is consistent with the metric (9) only when $\tilde{R}_{\alpha\beta}(y) \propto \tilde{g}_{\alpha\beta}(y)$, as seen from (14) and (19). We shall write with a nonnegative constant k,

$$\tilde{R}_{\alpha\beta}(y) = k\, \tilde{g}_{\alpha\beta}(y). \tag{20}$$

A more convenient set of equations are derived by eliminating $\sigma$ from (16) $\sim$ (19) and writing differential equations for b, the pressure $p_1$ and a quantity defined by

$$u \equiv \frac{1}{2} r\, (1 - e^{-\omega}). \tag{21}$$

In the ordinary relativity the Schwarzschild solution yields a constant u which coincides with half the gravitational radius. The equation for the size b(r) is obtained from (18) and (19) by eliminating $\sigma' - \omega'$ and $e^{-\omega}$ in (13) and (14) with (21),

$$(1- \frac{2u}{r})\, \{\frac{b''}{b} - (\frac{b'}{b})^2\} + \frac{2}{r}(1- \frac{u}{r})\frac{b'}{b} - k\, b^{-2}$$

$$= -\kappa^2 (d+2)^{-1} [\rho - 3p_1 + 2p_2 - r\frac{b'}{b}\{\rho + (d-1)p_1 - dp_2\}] . \tag{22}$$

The equation for u is obtained by eliminating $\sigma$ from (11) $\sim$ (13) and then rewriting the equation for $\omega$ in terms of u. One finds that

$$u' - \frac{d}{2}r(r-2u)\{\frac{1}{2}(1 + \frac{d}{2}\, r\, \frac{b'}{b})^{-1} \frac{b''}{b} + \frac{1}{r}\frac{b'}{b}\}$$

$$= \frac{\kappa^2}{2}\, r^2\rho - \frac{\kappa^2}{4}\, r^2 d(d+2)^{-1}(\rho-3p_1+2p_2) - \frac{\kappa^2}{8} r^3 d\frac{b'}{b}(1 + \frac{d}{2}\, r\, \frac{b'}{b})^{-1}(\rho+p_1). \tag{23}$$

To derive the last fundamental equation for the pressure gradient, one notes the energy-momentum conservation, $T_{A\ ;B}^{\ B} = 0$;

$$p_1' + \frac{1}{2}(\rho + p_1)\sigma' - d(p_2 - p_1)\, \frac{b'}{b} = 0. \tag{24}$$

By eliminating $\sigma'$ in this equation with (11) $\sim$ (13), one finds that

$$-p_1' + d\frac{b'}{b}(p_2 - p_1) = \frac{1}{2}(\rho + p_1)\, [\frac{2u}{r(r-2u)} + \kappa^2\frac{r^2}{r-2u}\, p_1$$

$$+ \kappa^2\, \frac{r^2}{r-2u}\, \frac{d}{2}(d+2)^{-1}(\rho - 3p_1 + 2p_2) - \frac{\kappa^2}{4}\, \frac{r^3}{r-2u}\, d\, \frac{b'}{b}\, (1 + \frac{d}{2}\, r\, \frac{b'}{b})^{-1}(\rho+p_1)$$

$$- d\frac{b'}{b} + (1 + \frac{d}{2}\, r\, \frac{b'}{b})^{-1}\, \frac{d}{2}\, r\, \frac{b''}{b}] . \tag{25}$$

The three equations, (22), (23) and (25), thus obtained are fundamental equations in spherically symmetric cases when they are supplemented by an equation of state, $p_1 = p_1(\rho)$ and $p_2 = p_2(\rho)$. The boundary condition for this set of equations is that asymptotically at $r = \infty$, $b(r)$ approaches a constant value $b_o$ in the ground state and the four dimensional metric $g_{ij}(x)$ approaches the flat metric, hence $\omega \to 0$ and $\sigma \to 0$. When the extra space does not exist, one can set, $d = 0$. In this case the basic equations, (23) and (25), reduce to the well known Tolman-Oppenheimer-Volkoff equation[10].

We now turn to vacuum solution when the extra compact space is Ricci-flat; $\tilde{R}_{\alpha\beta} = 0$, hence $k = 0$ in (20). Examples of the Ricci-flat space are a d-dimensional torus with $\tilde{R}_{\alpha\beta\gamma\delta} = 0$ and Calabi-Yau space with $\tilde{R}_{\alpha\beta\gamma\delta}\tilde{R}^{\alpha\beta\gamma\delta} \neq 0$. Vacuum equation with a Ricci-flat extra space is given by

$$(1 - \frac{2u}{r}) \{\frac{b''}{b} - (\frac{b'}{b})^2\} + \frac{2}{r}(1 - \frac{u}{r})\frac{b'}{b} = 0 \,, \tag{26}$$

$$u' - \frac{d}{2}r(r - 2u)(1 + \frac{d}{2}\, r\, \frac{b'}{b})^{-1} \frac{b'}{b}\{\frac{1}{2}(d+1)\frac{b'}{b} - \frac{u}{r(r-2u)}\} = 0. \tag{27}$$

The usual Schwarzschild solution follows with $b$ = const., and $u$ = const., which obviously obeys (26) and (27). One can extend this solution by assuming that $u = u_o$ (const.) alone. From (26)

$$\frac{b'}{b} = A\{r(r - 2u_o)\}^{-1}, \; A = \text{const.}$$

This is consistent with (27) only if $A = (d+1)^{-1}2u_o$. Thus, one finds a family of vacuum solutions given by

$$b = b_o(1 - \frac{2u_o}{r})^{1/(d+1)} \,, \tag{28}$$

$$e^{\omega} = (1 - \frac{2u_o}{r})^{-1} \,, \tag{29}$$

$$e^{\sigma} = (1 - \frac{2u_o}{r})^{-(d-1)/(d+1)} \tag{30}$$

The last equation (30) was obtained from

$$\sigma' + \omega' + \frac{d}{2}\, r\, \frac{b'}{b}(\sigma' + \omega') - dr\, \frac{b''}{b} = 0,$$

which follows from $e^{\omega-\sigma}R_{oo} + R_{rr} = 0$ with (11) and (12). The solution given by (28) ∿ (30) yields a curvature singularity at $r = 2u_o$, which appears in $R_{ABCD}R^{ABCD}$.

A new vacuum solution given by (28) ∿ (30) shows a peculiar behavior even far away from singular points at $r = 2u_o$. The Newtonian potential usually given by the metric component $g_{oo}$ yields

$$\frac{1}{2}(g_{oo} - 1) \sim \frac{\sigma}{2} \sim u_o(d-1)(d+1)^{-1}\frac{1}{r}\ . \qquad (31)$$

With $d > 1$, this yields a repulsive force, because the energy, defined in the usual way by the spatial component (29) at great distances, is $u_o/G$. The basic reason for this peculiarity is traced back to the Ricci-flatness of the extra manifold.

What happens when a positive curvature k is present? In this case the right hand side of the equation (26) for b is modified by a term, $kb^{-2} + V(b)$. For instance, when compactification is achieved by condensation of gauge[11] or antisymmetric tensor fields[12], or by the Casimir pressure[13],

$$kb^{-2} + V(b) = -k_1b^{-n} + kb^{-2} - k_2, \quad k_1 > 0,\ k_2 > 0,\ n > 2. \qquad (32)$$

This has a zero at $b = b_o$, and $kb^{-2} + V(b) \sim \mu^2 b_o^{-1}(b-b_o)$ around $b_o$. The second zero of (32) at $b > b_o$ does not correspond to a stable configuration. For the asymptotically constant case of $b \to b_o$, deviation of the size b(r) from $b_o$ is then cut off, like $b(r) - b_o \sim e^{-\mu r}/r$. The cut off length $\mu^{-1}$ is usually of order the Planck length, but in superstring theories it can be of order $m^{-1}$ with m the scale of supersymmetry breaking. How precisely the would-be long range force is cut off in superstring theories is not known at the moment.

It might be instructive to understand the peculiar vacuum solution in the four dimensional language. After the Weyl rescaling the scale factor essentially behaves as a massless scalar $\phi$ (∿ ln b). In this new metric the equivalence principle (equality of the inertial and the gravitational masses) is satisfied, but there is an extra long range force due to the scalar. At great distances,

$$\sigma \sim - \omega \sim - (d+1)^{-1} \frac{2u_o}{r} .$$

$$\phi \sim - \{\frac{1}{2}d(d+2)v_d\}^{1/2}\kappa^{-1}(d+1)^{-1} \frac{2u_o}{r} .$$

Taking into account an unambiguous magnitude of scalar coupling to matter, one precisely finds the previous potential (31). This analysis follows that of Gross and Perry[14)], who found similar peculiar solutions in five dimensional Kaluza-Klein theory in which the internal space is necessarily Ricci-flat.

In the same exploratory spirit, we shall seek a solution which has a uniform energy density $\rho$ and a constant size b (different from the vacuum value of $b_o$) within a radius R. In general, one can not find a solution with this ansatz, but remarkably a solution exists if the equation of state obeys

$$p_2 = \frac{3}{2} p_1. \tag{33}$$

We do not know what kind of matter this corresponds to, but since our intention here is to demonstrate peculiar features of extra dimensions, we shall not be bothered by this question. With the relation (33), the three basic equations, (22), (23) and (25), are as follows,

$$-k_1 b^{-n} + k b^{-2} - k_2 = \kappa^2(d+2)^{-1}\rho , \tag{34}$$

$$u' = \frac{1}{4}(d+2)^{-1}(d+4)\kappa^2 r^2 \rho + r^2(- \frac{d}{8} k_1 b^{-n} + \frac{1}{2} k_2) , \tag{35}$$

$$-p_1' = \{r(r-2u)\}^{-1}(p_1+\rho)[u + \frac{\kappa^2}{2}r^3\{p_1 + \frac{d}{2}(d+2)^{-1}\rho\}$$
$$+ r^3 (\frac{d}{8} k_1 b^{-n} - \frac{1}{2}k_2) ]. \tag{36}$$

For definiteness, we have considered the Casimir model[13)] of compactification in which the exponent n = d + 4 and the Casimir stress obeys a relation, $-p_1 = \rho = dp_2/4$. A model with the gauge field condensate can be treated in similar fashions.

From (35) it is clear that $u = \frac{1}{2}Cr^3$ when $r < R$, hence

$$-\{(p_1+\rho)(p_1+B)\}^{-1}p_1' = \frac{\kappa^2}{2}r(1-Cr^2)^{-1} , \tag{37}$$

$$B = \frac{2}{3}(d+2)^{-1}(d+1)\rho + \frac{4}{3}\kappa^{-2}(\frac{d}{8}k_1 b^{-n} - \frac{1}{2}k_2) , \tag{38}$$

$$C = \frac{\kappa^2}{2}(\rho - B). \tag{39}$$

The differential equation (37) is easily integrated with the boundary condition of $p_1(R) = 0$;

$$p_1 = [\rho(1-CR^2)^{1/2} - B(1-Cr^2)^{1/2}]^{-1}[(1-Cr^2)^{1/2} - (1-CR^2)^{1/2}]\rho B, \quad r < R. \tag{40}$$

From a condition that the pressure remains finite for $0 < r < R$, one finds that

$$(1 - CR^2)^{1/2} > \frac{B}{\rho} . \tag{41}$$

Metric components are computed from (21) and (24) to yield

$$e^{\omega} = (1 - Cr^2)^{-1} , \tag{42}$$

$$e^{\sigma} = (\rho - B)^{-2}[\rho(1 - CR^2)^{1/2} - B(1 - Cr^2)^{1/2}]^2. \tag{43}$$

It is interesting to compare this solution with the corresponding solution[9)] in general relativity. In the case of the ordinary four dimensional relativity the parameter $\rho/B$ is fixed to be equal to 3, otherwise solutions, (40), (42) and (43), are exactly the same as in our case. This parameter is physically important because it determines the red shift of light emitted form the surface at $r = R$. The red shift factor $z$ is given by[9)]

$$z = [(g_{oo})^{-1/2} - 1]_{r=R} = (1 - CR^2)^{-1/2} - 1 < (\rho/B) - 1. \tag{44}$$

Namely, the maximum red shift is limited by $\rho/B - 1$, while for uniform stars in the ordinary relativity it is 2. In our case this parameter is constrained by two equations, (34) and (38).

To analyze this restriction, let us recall that the ground state value $b_o$ of the compactified size must be consistent with the vacuum energy of $\rho = 0$ and with the zero effective cosmological constant in four dimensions. Thus one may write

$$- k_1(b^{-n} - b_o^{-n}) + k(b^{-2} - b_o^{-2}) = \kappa^2(d+2)^{-1}\rho \quad , \tag{45}$$

$$B = \frac{2}{3}(d+2)^{-1}(d+1)\rho + \frac{d}{6}\kappa^{-2}k_1(b^{-n} - b_o^{-n}) \quad , \tag{46}$$

$$b_o = [\frac{1}{4}(d+4)\frac{k_1}{k}]^{1/(n-2)} \quad . \tag{47}$$

With a fixed $\rho$, b given by a solution of eq. (45) gradually increases from $b_o$ as $\rho$ increases. The value b then reaches $b_c$ at $\rho = \rho_c$, above which b becomes complex. $\rho/B$ monotonically increases as $\rho$ rises, hence the maximum value of $\rho/B$ can be computed at $\rho = \rho_c$ and $b = b_c$. By simple calculations with $n = d + 4$ (Casimir model),

$$b_c = 2^{1/(d+2)}b_o \quad , \tag{48}$$

$$\rho_c = \kappa^{-2}b_o^{-2}k(d+4)^{-1}(d+2)[(d+2)2^{-2/(d+2)} - d] \quad , \tag{49}$$

$$(\frac{\rho}{B})_{max} = \frac{3}{2}(d+2)\{d + 2 - d\ 2^{2/(d+2)}\}[(d+2)\{d + 1 - d\ 2^{2/(d+2)}\} + \frac{d}{2}]^{-1} \quad . \tag{50}$$

The maximum red shift given by $(\rho/B)_{max}-1$ thus depends on the dimension via(50). Numerically,

$z_{max}$ = 2.658 (d = 1), 3.169 (d = 2), 3.575 (d = 3), 3.905 (d = 4), 4.179 (d = 5), 4.410 (d = 6), 4.607 (d = 7), ... , 7.096 (d = ∞)

At d = ∞, the analytic expression is as follows,

$$z_{max} = (3 - 2\ \ell n\ 2)/(3 - 4\ \ell n\ 2).$$

We have also computed the maximum surface red shift of uniform stars in the gauge condensate model[11)] in which the condensate stress obeys a relation, $-p_1 = \rho = -p_2 d/(d-4)$. With the same equation of state (33), the maximum red shift is

$$z_{max} = \frac{3}{4}d + 2 \ .$$

In this case the red shift increases from 3.5 ($d = 2$) to infinity ($d = \infty$) as d increases. Thus, the surface gravity can exceed the bound in the four dimensional relativity, but its value is model dependent.

Exact solutions discovered and analyzed so far serve to illustrate peculiar features associated with extra dimensions, but their relevance to realistic situations such as gravitational collapse is dubious. In the last section of this paper we shall outline a more realistic approach to explore extra dimensions. A good starting point in the study of extremely dense objects is the Tolman solution[10)] at infinite central density;

$$\rho = \rho_o \equiv \frac{3}{7}\kappa^{-2} r^{-2} , \qquad u = u_o \equiv \frac{3}{14} r ,$$

$$b = b_o , \qquad p_1 = \frac{1}{3}\rho, \qquad p_2 = 0 , \tag{51}$$

This solution is assumed to approximately describe the state of a dense object in an intermediate region, say $r_1 < r < r_2$. Outside this region, $r > r_2$, the equation of state, $p_1 = \rho/3$, for massless particles will be replaced by a more realistic nonrelativistic equation and corresponding solution to the stellar differential equation, (23) and (25) with $b = b_o$ and $\rho - 3p_1 + 2p_2 = 0$ required by consistency with (22), becomes much more complicated unlike (51). Inside this region, $r < r_1$, heavy excitation modes in extra dimensions will contribute and the size b will deviate from the ground state value $b_o$.

One can make an analysis in a linear perturbation theory when deviation from the Tolman solution (51) is small. By defining small quantities,

$$b = b_o(1+\varepsilon_b),\ u = u_o+\delta u,\ \rho = \rho_o+\delta\rho,\ p_1 = \frac{1}{3}\rho_o+\delta p_1,\ p_2 = \delta p_2,$$

one derives equations for these,

$$\varepsilon_b'' + \frac{5}{2}r^{-1}\varepsilon_b' - \mu^2\varepsilon_b = -\frac{7}{4}\kappa^2(d+2)^{-1}(\delta\rho-3\delta p_1+2\delta p_2) , \tag{52}$$

$$\delta u' + \frac{d}{7}r\varepsilon_b' + \tilde{\mu}^2 r^2\varepsilon_b = \frac{\kappa^2}{2}r^2(d+2)^{-1}(2\delta\rho + 3d\delta p_1 - 2d\delta p_2) , \tag{53}$$

$$- \delta p_1' + \frac{4}{7}d(\kappa r)^{-2}\varepsilon_b' - \kappa^{-2}\tilde{\mu}^2 r^{-1}\varepsilon_b - 2\kappa^{-2}r^{-4}\delta u = \frac{1}{2r}(\delta\rho + 2\delta p_1). \qquad (54)$$

Here $\mu^2$ and $\tilde{\mu}^2$ are model dependent factors of order, $b_o^{-2}$ or $m^2$ (m = mass scale of supersymmetry breaking). Deviation from the Tolman solution presumably starts when the energy density becomes of order $b_o^{-d-4}$, namely roughly at the scale of the Planck length. Away from this region perturbation damps exponentially. For instance, in the case of $\varepsilon_b$ it falls off like $r^{-5/4}e^{-\mu r}$ with $r > \mu^{-1}$, as seen from (52). If $m << b_o^{-1}$, there may be a region of a power law fall off between $b_o$ and $m^{-1}$. This region of $b_o < r < m^{-1}$ is most interesting from the point of exploring effects of extra dimensions. On the other hand, theoretically it is least known, primarily because dynamics of supersymmetry breaking is not well understood.

In summary, we have demonstrated two peculiar exact solutions in which the scale factor of extra dimensions may vary with three dimensional distance. The vacuum solution does not respect the Birkoff theorem, hence uniqueness of the black hole solution. The solution with a uniform density violates the bound on the surface gravity, hence on the maximum red shift from stars. These solutions seem to illustrate a variety of new phenomena that may occur in theories with extra dimensions.

References

1. M.B. Green and J.H. Schwartz, Phys. Lett. 149B (1984) 117; E. Witten, Phys. Lett. 149B (1984) 351; D.J. Gross, J. Harvey, E. Martinec and R. Rohm, Phys. Rev. Lett. 52 (1985) 502.
2. E. Witten, Princeton preprint (1983).
3. For a review, J.H. Schwartz, Phys. Rep. 89 (1982) 223.
4. For a review, M. Yoshimura, "New Directions in Kaluza-Klein Cosmology" in Proceedings of Takayama Workshop on Toward Unification and its Verification, ed. by Y. Kazama and T. Koikawa, KEK report 85-4 (1985).
5. M. Yoshimura, "Cosmology with Ricci-flat Compactification", KEK-TH-114.
6. P. Candelas, G.T. Horowitz, A. Strominger and E. Witten, Nucl. Phys. B258 (1985) 46.
7. D.J. Gross, J. Harvey, E. Martinec and R. Rohm, "Heterotic String Theory II. The Interacting Heterotic String", Princeton preprint (1985); C.G. Callan, E.J. Martinec, M.J. Perry and D. Friedan, "Strings in Background Fields", Princeton preprint (1985); A. Sen, Phys. Rev. Lett. 55 (1985) 1846.
8. M. Dine, R. Rohm, N. Seiberg and E. Witten, Phys. Lett. 156B (1985) 55; E. Cohen, J. Ellis, C. Gomez and D.V. Nanopoulos, Phys. Lett. 160B (1985) 62.
9. For example, S. Weinberg, "Gravitation and Cosmology", (J. Wiley, New York, 1972).
10. J.R. Oppenheimer and G.M. Volkoff, Phys. Rev. 55 (1939) 374; R.C. Tolman, "Relativity, Thermodynamics and Cosmology", (Oxford, 1934).
11. S. Randjbar-Daemi, A. Salam and J. Strathdee, Nucl. Phys. B214 (1983) 491.

12. P.G.O. Freund and M.A. Rubin, Phys. Lett. 97B (1980) 233.

13. T. Appelquist and A. Chodos, Phys. Rev. Lett. 50 (1983) 141; P. Candelas and S. Weinberg, Nucl. Phys. B237 (1984) 397.

14. D.J. Gross and M.J. Perry, Nucl. Phys. B226 (1983) 29.

# ON THE GRAVITATIONAL POTENTIAL FOR MANY-BODY SYSTEM

Tadayuki Ohta
Department of Physics,
Miyagi University of Education,
Aoba-Aramaki, 980 Sendai, Japan

## § 1 OUTLINE OF THE PROBLEM

We are now in the new era of astrophysical observations. And a wide-ranging interest in higher order gravitational effects to the order of Post-Post-Newtonian(PPN) approximation has been raised, especially in connection with the motion of the heavy binary stars. The gravitational potential for many-body system to this order was first derived by us more than ten years ago. We review the three different methods used to obtain the potential and discuss one of the applications of the methods to the problem including electromagnetic field.

Except several attempts in the case of two-body system, for these ten years, there has been no other paper than ours to give the explicit expression to the potential for many-body system. Recently our result was reconfirmed by T. Damour and G. Schafer who followed faithfully the calculation in one of our methods. They, however, do not notice the importance of the other two methods and have misunderstanding about our view to approach the problem. We will touch on this point in the following.

## § 2 FOKKER ACTION

We published four papers[1)~4)] on higher order gravitational potential. ( These are referred to as [I], [II], [III] and [IV]. ) The first method we used is based on the Fokker action principle in which the equations of motion of the bodies are given by the variation

$$\delta \int_{x^{\circ\prime}}^{x^{\circ\prime\prime}} L^{*}\, dx^{\circ} = 0, \tag{1}$$

where the Lagrangian $L^{*}$ is

$$L^{*} = \int dx^{3}\{ - \sum_{a} m_{a}\, c^{2}\, \delta( \vec{x} - \vec{z}_{a} ) \sqrt{- g_{\mu\nu} \frac{dz_{a}^{\mu}}{dx^{\circ}} \frac{dz_{a}^{\nu}}{dx^{\circ}}}$$

$$- \frac{c^{4}}{16\pi G} \sqrt{-g}\, g^{\mu\nu} ( \Gamma^{\rho}_{\mu\nu} \Gamma^{\lambda}_{\rho\lambda} - \Gamma^{\lambda}_{\mu\rho} \Gamma^{\rho}_{\nu\lambda} ) \} . \tag{2}$$

Here, G, $g_{\mu\nu}$, g, $\Gamma^{\rho}_{\mu\nu}$, $m_a$ and $\vec{z}_a$ are Newton's gravitational constant, metric tensor, det($g_{\mu\nu}$), Christoffel's symbol, the rest mass and the coordinate of a-th body, respectively, and $z^{o}_{a,o}= 1$. In the expression (2), $g_{\mu\nu}$ and $\Gamma^{\rho}_{\mu\nu}$ should be given as the functions of the coordinates ($\vec{z}_a$'s) and the velocities ($\dot{\vec{z}}_a$'s) of the bodies. The variation (1) leads to the correct equations of motion in general relativity. The formal proof of this action principle is given in the textbook by L.Infeld and J. Plebanski.[5)]

Before calculating the Lagrangian we have to know the explicit expression of the metric tensor $g_{\mu\nu}$ up to the sixth order of $c^{-1}$ (c: velocity of light ). In a process of solving Einstein's equation, we found that in the coordinate conditions used in the popular literatures the metric tensor was divergent at spatial infinity ($r \to \infty$). When the metric tensor is not flat at spatial infinity, the Fokker action principle no longer holds because of the divergence of one surface integral in the action. So, we tried to look for the coordinate condition under which the metric tensor was Minkowskian at spatial infinity. We found a class of coordinate conditions which led to well-behaved metric tensors.[1)]

We encountered, however, another serious difficulty. When the Lagrangian (2) is evaluated by substituting the explicit expression of the metric tensor in it, the terms with acceleration appear in the Lagrangian $L^*$. Such an acceleration dependent Lagrangian is not only beyond the ordinary framework of dynamical theory (Hamilton's principle), but also a generalized Euler equation derived from it, is no longer Infeld's equation which is equivalent to the geodesic equations of motion in general relativity. Even if the action principle itself could be improved to incorporate with the acceleration terms, it would be meaningless to our problem.

Next we made a conjecture on what happened when we rewrote the acceleration terms in the original Lagrangian $L^*(\vec{z}_a, \dot{\vec{z}}_a, \ddot{\vec{z}}_a)$ with the use of the equations of motion in the Newtonian approximation $\ddot{\vec{z}}_a = -\Sigma \dfrac{Gm_b}{r_{ab}^2} \vec{n}_{ab}$, $r_{ab} = |\vec{z}_a - \vec{z}_b|$ and $\vec{n}_{ab} = (\vec{z}_a - \vec{z}_b)/|\vec{z}_a - \vec{z}_b|$. The equations of motion derived from the modified Lagrangian $L(\vec{z}_a, \dot{\vec{z}}_a)$ thus obtained, are not equivalent to the generalized equations derived from the original Lagrangian $L^*$.

At this stage we could not judge whether the modified Lagrangian had a physical meaning, or not. We found, however, that the Hamiltonian derived from the modified Lagrangian was identical with the Hamiltonian obtained in another formalism described in the next section.

## §3 HAMILTONIAN IN THE CANONICAL FORMALISM

An essential point of the problem is considered to be in the relation between the coordinate condition and the potential. So we chose the canonical formalism by ADM[6)] as another method to get the potential. This formalism is best suited to obtain the Hamiltonian which is given as a simple three space integral of some part of the metric tensor,

$$H = -\int dx^3 \, \Delta h^T \, , \tag{3}$$

$$h^T = h_{ii} - \frac{1}{\Delta} h_{ij,ij} \, , \quad h_{ij} = g_{ij} - \delta_{ij} \, , \; (i,j=1,2,3). \tag{4}$$

In this formalism, of course, the metric tensor must be Minkowskian at spatial infinity.

We investigated a wider class of coordinate conditions and found that in a certain class of coordinate conditions Hamiltonian contained only the coordinates and velocities of the bodies. In the most simple condition we obtained the explicit expression of the Hamiltonian to the PPN order of approximation. This result was given in the paper [III] and was found to be identical with the Hamiltonian derived from the modified Lagrangian $L(\vec{z}_a, \dot{\vec{z}}_a)$ in the method in §2.

In the section 5 of the paper [III], we examined in detail how this coincidence occurred. We found

$$H^{*} = H(\vec{z}_a, \vec{p}_a) - \partial_0 A(\vec{z}_a, \vec{p}_a) , \qquad (5)$$

where H is the Hamiltonian given in the coordinate system chosen in [III] and $H^*$ is the one formally evaluated by using the metric tensor in the paper [I]. The difference between H and $H^*$ is a total time derivative of a function A and it may seem to have no physical effect. However, the actual expressions of A and $H^*$ include accelerations of the bodies. Then $H^*$ is beyond the framework of the standard Hamiltonian formalism. The Hamiltonian which is able to describe dynamics of physical system is $H(\vec{z}_a, \vec{p}_a)$ given in [III]. The Hamilton's equation by $H(\vec{z}_a, \vec{p}_a)$ is equivalent to Infeld's equation.

## § 4 POTENTIAL FROM S-MATRIX IN QUANTUM FIELD THEORY

In order to confirm the result obtained in [III], we tried quite a different approach to get the classical potential. That is the S-matrix method.

As the first step we calculate S-matrix element for a scattering of n-particles in quantized field theory. Next we sum up the contributions of all tree diagrams. Finally, by fourier transformation of the sum in the limit of $h \to 0$, we get a classical n-particle potential. There is a general criterion that the result in a quantized theory should coincide in the limit of $h \to 0$ with the classical one. Since we need the contributions which survive in the limit $h \to 0$, we have only to pick up tree diagrams. And we do not need to touch on a renormalization problem.

In order to get the potential to the PPN order of approximation, we had to calculate the matrix elements for scatterings of two, three, and four-particles. The framework of the quantized field theory we chose is the system of scalar particles and gravitons. We treated the graviton field in a conventional quantization of a field Lagrangian. Namely, we define the graviton field $h_{\alpha\beta}$ and the free Lagrangian $L_0$ as

$$g_{\alpha\beta} = \delta_{\alpha\beta} + \kappa h_{\alpha\beta} \quad , \; (\kappa^2 = 32\pi G) \tag{6}$$

and

$$L_0 = -\frac{1}{2} h_{\beta\gamma,\alpha} h_{\beta\gamma,\alpha} + \frac{1}{2} h_{\beta\beta,\alpha} h_{\gamma\gamma,\alpha} + h_{\beta\gamma,\alpha} h_{\alpha\gamma,\beta} - h_{\alpha\beta,\alpha} h_{\gamma\gamma,\beta} \quad . \tag{7}$$

All higher terms of $h_{\alpha\beta}$ in the total Lagrangian are treated as interactions. We defined the graviton propagator $X_{\alpha\beta,\gamma\delta}(k)$ in a preferred gauge and calculated S-matrix elements in all tree diagrams. Examples of the diagrams are shown in Fig.1.

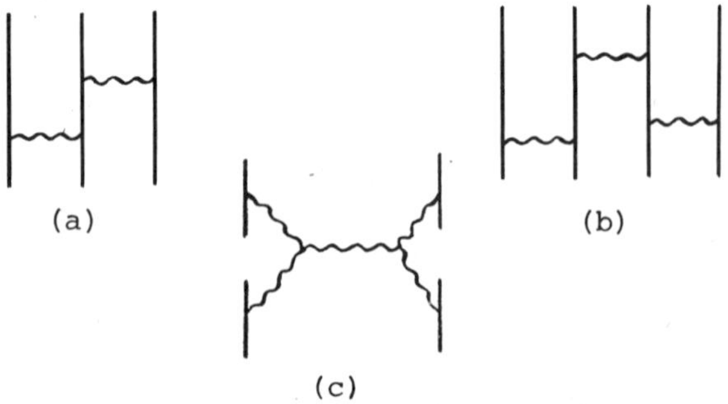


Fig.1. Examples of the tree diagrams in three and four-particle scattering. Straight lines represent scalar particles and wavy lines represent exchanged gravitons.

The potential to the Post-Newtonian approximation had already been calculated in this approach by Iwasaki[7], and Hiida and Okamura.[8] The result coincides with the standard one given by Einstein, Infeld and Hoffmann. In the paper [IV] we calculated the PPN potential, namely $G^3$, $G^2v^2$ and $Gv^4$ potentials.

In the calculation of the $G^3$ potential, we were faced with one difficulty that the result was different from that obtained in the canonical formalism. Later this difficulty was resolved by Yokoya et al..[9] They found that the origin of this discrepancy was in the process of the subtraction. For example, in order to get a three-body potential through the diagram (a), we have to subtract the contribution of the repetition of two-particle scattering by two-body potential.

In the PPN order, $G^3$-static potential is derived from the S-matrix element for four-particle scattering. Then in the diagram as (b), we need double subtractions. Yokoya et al.[9] pointed out that there remained one arbitrary parameter in the potential through the process of subtraction. It is caused when the energy of an internal particle is expressed with the physical quantities of external particles. The four-body static potential from the S-matrix is

$$(12-27\eta)\ V_A - \frac{3}{4} V_B + V^{TT} \quad , \tag{8}$$

where $V^{TT}$ is the contribution from the transverse-traceless part of the graviton in the diagram (c). The symbols $V_A$ and $V_B$ denote the potentials in the forms

$$\Sigma \frac{G^3 m_i m_j m_k m_l}{r_{ij} r_{ik} r_{il}} \qquad \text{and} \qquad \Sigma \frac{G^3 m_i m_j m_k m_l}{r_{ij} r_{jk} r_{kl}} \quad ,$$

respectively, where $r_{ij}$ is the distance between i-th and j-th particles. From a physical point of view, the parameter $\eta$ has to take the value 1/2. This point was discussed by one of our colleagues.[10)] In the paper [IV] we had chosen an unappropriate value $\eta = 0$. Now by performing the consistent subtraction in all orders of the diagrams, we can derive in the S-matrix approach the gravitational potential given in the paper [III].

The important point in this third method is as follows. In the calculation of the S-matrix element for particle scattering, physical quantities denoting the external particles are only the masses and momenta. The coordinates of the particles are introduced through the fourier integration on the momenta of internal gravitons. Then the final potential is expressed with the masses, momenta and coordinates of the particles. There is no place in which the acceleration of the particle enters. This method leads to the Hamiltonian which is a function of the coordinates and the momenta.

## § 5 RECONFIRMATION OF OUR RESULT AND THE EXPLICIT EXPRESSION OF TWO-BODY POTENTIAL

For these ten years, there was no paper than ours to obtain many-body gravitational potential in higher orders. Recently, T. Damour and G. Schafer[11)] have faithfully followed many cumbersome calculations we described in the paper [I] and [II]. They reconfirmed our result and quoted our Lagrangian as it was in our paper. They used our notations and symbols. It may give an incorrect impression to the readers, as if the result is their original work. Their work is a reconfirmation of ours. Furthermore, they misunderstand our consideration described above in §2 and do not touch intentionally on the most important paper of our series, namely the method in the canonical formalism explained in §3.

In the last part of the paper [II], we gave the explicit expression of the two-body Hamiltonian, in which there was one complicate integral $V^{TT}$, arising from the contribution of the transverse-traceless part of the graviton field. The coefficient of one of the integrals in $V^{TT}$ was not correct. This was pointed out by Damour and Schafer after checking many troublesome divergent integrals. We acknowledge their indication.

The correct form of the two-body Hamiltonian is

$$H(1,2) = m_1 + m_2 + \frac{\vec{p}_1^{\,2}}{2m_1} + \frac{\vec{p}_2^{\,2}}{2m_2} - \frac{1}{8}\{ m_1\left(\frac{\vec{p}_1^{\,2}}{m_1^2}\right)^2 + m_2\left(\frac{\vec{p}_2^{\,2}}{m_2^2}\right)^2 \}$$

$$+ \frac{1}{16}\{ m_1\left(\frac{\vec{p}_1^{\,2}}{m_1^2}\right)^3 + m_2\left(\frac{\vec{p}_2^{\,2}}{m_2^2}\right)^3 \} - \frac{Gm_1m_2}{r} + \frac{G^2}{2}\cdot \frac{m_1m_2(m_1+m_2)}{r^2}$$

$$- \frac{G^3}{4} \cdot \frac{m_1 m_2 (m_1^2 + m_2^2 + 5 m_1 m_2)}{r^3} + \frac{G^2}{4} \cdot \frac{m_1 m_2^2}{r^2} \left(10 \frac{\vec{p}_1^{\,2}}{m_1^2} + 19 \frac{\vec{p}_2^{\,2}}{m_2^2}\right)$$

$$+ \frac{G^2}{4} \cdot \frac{m_1^2 m_2}{r^2} \left(19 \frac{\vec{p}_1^{\,2}}{m_1^2} + 10 \frac{\vec{p}_2^{\,2}}{m_2^2}\right)$$

$$- \frac{G^2}{4} \cdot \frac{m_1 m_2 (m_1 + m_2)}{r^2} \left\{ 27 \frac{(\vec{p}_1 \cdot \vec{p}_2)}{m_1 m_2} + 6 \frac{(\vec{n} \cdot \vec{p}_1)(\vec{n} \cdot \vec{p}_2)}{m_1 m_2} \right\}$$

$$+ \frac{G}{8} \cdot \frac{m_1 m_2}{r} \left[ -12 \left( \frac{\vec{p}_1^{\,2}}{m_1^2} + \frac{\vec{p}_2^{\,2}}{m_2^2} \right) + 28 \frac{(\vec{p}_1 \cdot \vec{p}_2)}{m_1 m_2} + 4 \frac{(\vec{n} \cdot \vec{p}_1)(\vec{n} \cdot \vec{p}_2)}{m_1 m_2} \right.$$

$$+ 5 \left\{ \left( \frac{\vec{p}_1^{\,2}}{m_1^2} \right)^2 + \left( \frac{\vec{p}_2^{\,2}}{m_2^2} \right)^2 \right\} - 11 \frac{\vec{p}_1^{\,2} \vec{p}_2^{\,2}}{m_1^2 m_2^2} - 2 \frac{(\vec{p}_1 \cdot \vec{p}_2)^2}{m_1^2 m_2^2}$$

$$+ 5 \frac{\vec{p}_1^{\,2} (\vec{n} \cdot \vec{p}_2)^2 + (\vec{n} \cdot \vec{p}_1)^2 \vec{p}_2^{\,2}}{m_1^2 m_2^2} - 12 \frac{(\vec{p}_1 \cdot \vec{p}_2)(\vec{n} \cdot \vec{p}_1)(\vec{n} \cdot \vec{p}_2)}{m_1^2 m_2^2}$$

$$\left. - 3 \frac{(\vec{n} \cdot \vec{p}_1)^2 (\vec{n} \cdot \vec{p}_2)^2}{m_1^2 m_2^2} \right], \tag{9}$$

where $\vec{n} = \vec{n}_{12}$ .

## § 6 STATIC BALANCE OF CHARGED TWO-BODY SYSTEM

The effects of the gravitational potential in the PPN order to physical phenomena seemed very small. However, as the techniques on astrophysical observations make rapid progress, these effects will become measurable in a near future, for example, in the motion of binary stars. As the applications of many-body gravitational effects, we have investigated several problems. Here we briefly state one of them, the problem of static balance in general relativity.

The problem of the static balance has originated in the search for exact solutions to Einstein's equation. It is evident that there is no exact solution in the case of a gravitationally interacting two-body system. However, in a rather peculiar situation that each body has electric charge, there might be an exactly balanced solution. And this was discovered by S. D. Majumdar[12] and A. Papapetrou[13] under the Weyl's assumption. Since then, there have been many discussions on the uniqueness of static balance, because the condition given by them is $e_i = \pm\sqrt{4\pi G}\cdot m_i$ (i=1,2) and is more strict than the balance condition in the classical theory, $e_1 e_2/4\pi = Gm_1 m_2$. Why is the balance condition in general relativity different from that in the classical theory ? This is a naive question.

In 1977, B. M. Barker and R. F. O'Connell[14] studied the electro-gravitational potential to the Post-Newtonian order and pointed another possibility of static balance, $e_i=\sqrt{4\pi G m_1 m_2}$ (i=1,2). We investigated this question through the electro-gravitational potential derived in the two methods described in §4 and §5. From the discussion on the potential up to the PPN order, we concluded that $e_i = \pm\sqrt{4\pi G}\cdot m_i$ was the only condition of static balance in general relativity.[15]

Later, W. B. Bonnor [16] proposed the conjecture that the balance condition in general relativity would be identical with that in Newtonian mechanics. He actually solved Einstein-Maxwell equation in empty space and obtained the solution of metric tensor to the PPN order of approximation. He deduced the balance condition $e_1e_2/4\pi = Gm_1m_2$ from the requirement of vanishing of a strut singularity which existed on a straight line connecting the positions of two particles. There appeared, however, one peculiar term in $g_{\mu\nu}$ which would destroy the basis of his conjecture. He considered that the term arose from unnecessary multipole contributions.

We tackled the same problem in a similar method. We dealt with Einstein-Maxwell equation with a source term of dipole and showed that the peculiar term really existed in the system of two particles. We concluded that Bonnor's conjecture was not right, because the peculiar term came from dipole contribution.[17] The balance condition in general relativity is $e_i = \pm\sqrt{4\pi G}\cdot m_i$ (i=1,2).

ACKNOWLEDGEMENTS

It is my great pleasure that Professor G. Takeda has seen his 60th birthday come round and is now more active in the theoretical research of elementary particles. For a long period since my graduate student days, I have greatly indepted to his extensive guidance to the world of physics.

I also thank my colleagues, especially Professor T. Kimura, for many useful discussions.

REFERENCES

1) T. Ohta, H. Okamura, T. Kimura and K. Hiida, Prog. Theor. Phys. 50 (1973), 492. (referred to as [I]).

2) T. Ohta, H. Okamura, T. Kimura and K. Hiida, Prog. Theor. Phys. 51 (1974), 1220. (referred to as [II]).

3) T. Ohta, H. Okamura, T. Kimura and K. Hiida, Prog. Theor. Phys. 51 (1974), 1598. (referred to as [III]).

4) H. Okamura, T. Ohta, T. Kimura and K. Hiida, Prog. Theor. Phys. 50 (1973), 2066. (referred to as [IV]).

5) L. Infeld and J. Plebanski, Motion and Relativity (Pergamon Press, 1960), Chapter IV.

6) R. Arnowitt, S. Deser and C. W. Misner, Phys. Rev. 116 (1959), 1322; 117 (1960), 1959; 118 (1960), 1100.

7) Y. Iwasaki, Lett. Nuovo Cim. 1 (1971), 783; Prog. Theor. Phys. 46 (1971), 1587.

8) K. Hiida and H. Okamura, Prog. Theor. Phys. 47 (1972),1743.

9) K. Yokoya, A. Shimizu, H. Kato and K. Hiida, Phys. Rev. D. 16 (1977), 2655.

10) T. Kimura, Prog. Theor. Phys. 57 (1977), 1437.

11) T. Damour and G. Schafer, General Relativity and Gravitation, 17 (1985), 879.

12) S. D. Majumdar, Phys. Rev. 72 (1947), 390.

13) A. Papapetrou, Proc. Roy. Irish Academy 51 (1947), 191.

14) B. M. Barker and R. F. O'Connell, Phys. Letters 61A (1977) 297; J. Math. Phys. 18 (1977), 1818.

15) T. Kimura and T. Ohta, Phys. Letters 63A (1977), 193.

16) W. B. Bonnor, Phys. Letters 83A (1981), 414.

17) T. Ohta and T. Kimura, Phys. Letters 90A (1982), 389; Prog. Theor. Phys. 68 (1982), 1175.

# A NEW GAUGE TRANSFORMATION FOR HIGHER-SPIN GAUGE FIELDS

Kazunari Shima
Laboratory of Physics
The Saitama Institute of Technology
Okabe-machi, Saitama 369-02, Japan

ABSTRACT

We study a gauge transformation which rotates simultaneously gauge fields with spin 5/2 and 3/2(in general spin s and s-1). A possible relevance to constructing consistent gravitational couplings of higher-spin gauge fields is investigated.

Gauge invariance plays essential roles to construct field theory of massless high-spin particles. The systematics of higher-spin gauge fields is studied extensively in ref.[1]. Furthermore, to overcome the advocated difficulties in gravitational interaction of higher-spin massless particles deeper understandings of the structure of gauge transformations are indispensable.
Superstring theory does have these higher-spin (massless) states[2] and $10 \leq N \leq 12$ supergravity needs such states.[3]

In this letter we survey a gauge transformation which rotates gauge fields with spin 5/2 and 3/2 simultaneously.
And we suggest that it may play an important role to construct consistent gravitational couplings of higher-spin($\geq$ 5/2) gauge fields.

In advance, to clarify our discussions we briefly review the arguments of ref.[1] for massless spin 5/2 case, for it is the highest fermionic spin contained in N = 10 supergravity which accomodates all observed particles as elementary ones in a minimal way. For further details see ref.[3].

From geometrical arguments, they obtain a field equation $Q_{\mu\nu}(\Psi_{\rho\sigma}) = 0$ for symmetric tensor-spinor $\Psi_{\rho\sigma}$ which has 40 components[1]. The field equation is invariant under gauge transformation $\delta\Psi_{\rho\sigma} = \partial_{\{\rho}\varepsilon_{\sigma\}}$, where $\varepsilon_{\mu}(x)$ is vector-spinor gauge parameter subject to constraint $\gamma^{\mu}\varepsilon_{\mu} = 0$, i.e. $\varepsilon_{\mu}(x)$ has 12 independent components. The count of physical degrees of freedom of $\Psi_{\mu\nu}$ in $Q_{\mu\nu} = 0$ is as follows :

(i) linear combinations of $Q_{\mu\nu}$, i.e. $Q_{0j} - \gamma_0\gamma^{i}Q_{ij} = 0$, where i and j are spatial indices comprise $2\cdot 2(2+1) = 12$ constraints which contain only spatial derivatives,

(ii) derivative free covariant gauge conditions $F_\mu(\Psi_{\rho\sigma}) = 0$ and $\gamma^\mu F_\mu(\Psi_{\rho\sigma}) = 0$ gives 12 independent conditions and

(iii) variation of $F_\mu(\Psi_{\rho\sigma}) = 0$ under $\delta\Psi_{\mu\nu}$, i.e. $\delta F_\mu = 0$ and $\gamma^\mu \delta F_\mu = 0$ allows regauge transformations which remove 12 degrees of freedom.

From (i), (ii) and (iii) the number of dynamical components of $\Psi_{\mu\nu}$ becomes $40 - 12 \cdot 3 = 4$ to descrive helicity $\pm 5/2$ states. Similar arguments hold for spin 3/2 gauge field $\Psi_\mu$ with 16 components, where field equation $Q_\mu(\Psi_\rho) = 0$ is invariant under $\delta\Psi_\mu = \partial_\mu \varepsilon$ and unconstrained gauge parameter $\varepsilon(x)$ has 4 independent components. After the similar arguments to (i),(ii) and (iii) the number of dynamical components of $\Psi_\mu$ becomes $16-4\cdot 3=4$ to descrive helicity $\pm 3/2$ states. Up to now gauge transformation for $\Psi_\mu$ and $\Psi_{\mu\nu}$ are considered separately.

However from above discussions we should notice the possibility that $\Psi_{\mu\nu}$ and $\Psi_\mu$ can be rotated simultaneously by using one unconstrained gauge parameter $\varepsilon_\mu(x)$ which has 16 independent components(12 for $\Psi_{\mu\nu}$ and 4 for $\Psi_\mu$ ). To see this we need the most general gauge invariant equation of motion for $\Psi_{\mu\nu}$, where unconstrained gauge parameter $\varepsilon_\mu(x)$ is allowed.

Such an equation is already found[4] and as follows(in ref.[4] we take $\sigma=1$.)

$$I_{\mu\nu} = \not\partial\Psi_{\mu\nu} + \frac{1}{4}\gamma_\mu \not\partial \gamma^\lambda \Psi_{\lambda\nu} + \frac{1}{4}\gamma_\nu \not\partial \gamma^\lambda \Psi_{\lambda\mu} - \frac{1}{4}\not\partial \Psi^\lambda{}_\lambda \eta_{\mu\nu} = 0, \quad (1)$$

which is invariant under

$$\delta\Psi_{\mu\nu} = \frac{1}{2}\not\partial(\gamma_\mu \varepsilon_\nu + \gamma_\nu \varepsilon_\mu) - \eta_{\mu\nu}\partial\cdot\varepsilon, \quad (2)$$

where all 16 components of $\varepsilon_\mu$ do not generate $\delta\Psi_{\mu\nu}$ as noted in ref.[4]. Remarkably we find that using (1) and (2) we can carry out the similar arguments to (i), (ii) and (iii) to count the number of independent components of $\Psi_{\mu\nu}$. For example, constraint (i) now becomes $I_{j0} + \gamma^i\gamma^0 I_{ij} - \gamma_j\gamma^i I_{i0} - \frac{1}{2}\gamma_j I_{00} = 0$, which produces 12 constraints as before. Then 12 components of $\varepsilon_\mu$ are used for $\Psi_{\mu\nu}$ and 4 are still left for $\Psi_\mu$. To see this explicitly, we write 4 components $\varepsilon(x)$ as follows

$$\varepsilon_\mu(x) = \kappa\{\, \partial_\mu \varepsilon(x) - p\gamma_\mu \not\partial \varepsilon(x) \,\}\ , \quad p \neq \frac{1}{4}\ , \tag{3}$$

where $\kappa$ is a constant with dimension -1 and $p$ is a dimensionless parameter. As seen later the expression (3) with derivatives and factor $\kappa$ (identified later with Planck length) are suggested from the discussions on the consistency of the gravitational interactions of spin 5/2 gauge field. Alternatively we can write (3) as follows

$$\varepsilon(x) = \frac{1}{\kappa(1-p)} \int D(x-y)\partial^\rho \varepsilon_\rho(y)dy, \quad p \neq 1 \qquad \text{for } \partial^\rho\varepsilon_\rho(y) \neq 0$$

and (4)

$$\varepsilon(x) \ \text{ is arbitrary,} \quad p = 1 \qquad \text{for } \partial^\rho\varepsilon_\rho(y) = 0\ ,$$

where $D(x-y)$ is a Green's function of the d'Alenbertian operator. These show that for any $\varepsilon_\mu(x)$, $\varepsilon(x)$ defined by (3) has 4 independent components.

Next we check the regauge conditions (ii) for $\varepsilon_\mu$. By using (2) regauge conditions now become as follows

$$\delta F_\mu = -2\not\partial\varepsilon_\mu + \partial_\mu\not\varepsilon + \gamma_\mu\partial\cdot\varepsilon - \frac{3}{4}\gamma_\mu\not\partial\not\varepsilon = 0, \qquad (5)$$

which is remarkably traceless for unconstrained $\varepsilon_\mu(x)$ and gives 12 regauge conditions to remove 12 components from $\Psi_{\mu\nu}$. The remaining 4 components are used for $\varepsilon(x)$ of (4), i.e. by (re)gauge transformations they remove 4 components from $\Psi_\mu$.

From above discussions it can be concluded that using unconstrained parameter $\varepsilon_\mu(x)$, $\Psi_{\mu\nu}(x)$ and $\Psi_\mu(x)$ are rotated simultaneously. And for any $\varepsilon_\mu(x)$, $\varepsilon(x)$ defined by (1) can be used as a gauge parameter for $\Psi_\mu(x)$. The above counting of independent components of $\Psi_\mu$ and $\Psi_{\mu\nu}$ are formally written as follows $(40+16)-16(=12+4)\cdot 3=4+4$. Our new gauge transformation can be easily generalized to higher-spin gauge fields with spin s and s-1, and perhaps to a multiplet s,s-1,s-2,··,1. Probably this suggests a certain geometrical multiplet structure between higher-spin gauge fields with spin s and s-1. Such structures may be seen in superstring theory.

Up to now our discussions are confined only to free field cases and our gauge transformation does not produce any new physical results at all. However , as sketched below our new gauge transformation may play an important role in gravitational interaction of higher-spin($\geq 5/2$) gauge fields.

As usual, we introduce gravitational interaction in a minimal way by replacing $\partial_\mu \to D_\mu$(covariant) in (1) and (2) with factor $\kappa$ . Following the standard arguments the inconsistency problem may be set up as the problem of the gauge invariance of (1) under (2) in their covariant forms.

Taking the variation of (1) we obtain terms like

$$\kappa^{-1}R_{\rho\mu}\Gamma^{\rho}\varepsilon_{\nu} \qquad \text{and} \qquad \kappa^{-1}R_{\rho\mu\nu}{}^{\sigma}\Gamma^{\rho}\varepsilon_{\sigma} \, , \tag{6}$$

where $\Gamma^{\rho}$ is a certain $\gamma$-matrices with index $\rho$ . The first term including Ricci tensor can be eliminated by graded Lie algebra extension of gauge transformation as in supergravity for spin 3/2 but the second term of (6) including uncontracted full Riemann tensor , which is absent for spin 3/2 , can not be eliminated . This is the advocated difficulty in gravitational interaction of higher-spin($\geq$5/2) massless fields.

However using above new gauge transformation and with the help of spin 3/2 gauge field we can eliminate Riemann tensor terms as follows. At first, we add to (1) the following nonminimal interaction term

$$\kappa R_{\rho\mu\sigma\nu}\Gamma^{\rho}\Psi^{\sigma} \, . \tag{7}$$

Under $\delta\Psi_{\mu} = \kappa^{-1}D_{\mu}\varepsilon(x)$ , (7) produces a term which cancels Riemann tensor terms in (6), where $\varepsilon(x)$ is rewritten by (3). However from Lagrangian theory, we should add to spin 3/2 part the unwanted terms corresponding to (7). In this article, we do not discuss these terms further but only to notice that all such unwanted Riemann tensor terms can be eliminated by introducing derivative and nonminimal interaction terms of $\Psi_{\mu}$ and $R_{\rho\mu\sigma\nu}$[5]. Such terms are higher order in $\kappa$ . For further details, see ref.[5]. So far, up to Ricci tensor terms we have prove the consistency of above interacting system, while Riemann tensor appears in the action. We expect that starting from these Lagrangian, constructing interaction terms order by order in $\kappa$ and embedding new gauge transformation in graded Lie algebra, we can construct an invariant whole action. Accordingly equation (3) should be modified (by higher order in $\kappa$ ). To see easier the possibility of the above simultaneous gauge transformation for only free fields, in stead of (3), we should take

$$\varepsilon(x) = \gamma^{\mu}\varepsilon_{\mu}(x). \tag{8}$$

Nonminimal interaction terms mentioned above may be derived from a certain new minimal interaction as is anomolous magnetic moment of electron in standard electroweak GUTs.

Finally, the author hopes that this tiny work can bear against the severe and warm critique of Professor Gyo Takeda, to whom he has been owing very much since his start of studying high energy physics, and becomes a tiny dedication to the 60th birthday of Professor Gyo Takeda. Is it Possible?

He also thanks Professors Yasuhisa Murai, Kenju Mori and Takeshi Shirafuji of Saitama University for their useful discussions and warm hospitality extended to him.

## REFERENCES

[1] B.de Wit and D.Z. Freedman, Phys.Rev.D21(1980)358.

[2] For a recent review dictionary, see Superstring(World Sci.1985), ed. by J. Schwarz.

[3] K. Shima, Z.Phys.C18(1983)25.

[4] F.A.Berend, J.W.van Holten, P.van Nieuwenhuisen, and B.de Wit,Phys.Lett.83B(1979)188, Nucl.Phys.B154(1979)267.

[5] K. Shima, preprint SIT-85-3,submitted to publication.
preprint SIT-85-4,in preparation.

# ON THE SCHRODINGER FIELD

Yasushi Takahashi
Theoretical Physics Institute
Department of Physics
University of Alberta
Edmonton, Alberta, T6G 2J1, Canada

## ABSTRACT

A brief but systematic discussion of the Schrödinger field is presented from the view point of quantized field theory. It is pointed out that the local momentum conservation equation is not of the usual continuity equation type when two-body potential interaction is present and nevertheless the total momentum is globally conserved. The Schrödinger equation can be cast into a multicomponent equation containing only first order derivatives, depending on its spin contents. In case of spin 1/2, the g-factor is shown to be 2 even in purely non-relativistic Schrödinger field, in contrast with the general belief that g=2 is a relativistic effect.

## 1. INTRODUCTION

The purpose of this article is to collect relevant and useful relations associated with the quantized Schrödinger field in the presence of the two-body potential and the electromagnetic interaction.

As is well-known, the Schrödinger field is one of the most fundamental objects in non-relativistic many-body theory. Nevertheless, a systematic presentation of the Schrödinger field appears very rarely in text-books of quantum field theory, though relevant relations and formulas useful in many-body theory can be collected from various books and research papers if necessary. In view of this situation, we shall discuss, in section 2, the Schrödinger field with the internal and external potentials. In particular expressions for various physical quantities will be given. The usual canonical quantization technique will be used with numbers of incidental remarks. In section 3, the Galilei transformation will be discussed.

Section 4 is devoted to deriving an alternative form of the Schrödinger field with spin 1/2, which contains only the first order derivatives. The alternative form of the Schrödinger equation can be used in section 5 to show that the magnetic g-factor is 2 for spin 1/2, in spite of the wide-spread belief that the relation g=2 is obtained only when the electron is treated relativistically by way of the Dirac equation. It is seen that what is important for g=2 is not the relativity theory but the equation of the first order. This fact was emphasized by Lévy-Leblond in 1967[1].

One of the reasons why I have chosen this topic for the celebration of Prof. Takeda's 60th birhtday is that in the early stages of my research career, Prof. Takeda taught me to reinvestigate properly and thoroughly any argument usually taken for granted. A beautiful example of Prof. Takeda's research work along such an idea is seen in a series of papers[2], with Prof. Koba, where the Compton scattering, electron-electron scattering and general scattering processes are carefully calculated after Dancoff's work[3], and this analysis played the most essential role in establishing the theory of renormalization.

It is hoped that the presentation of this short article provides some readers with a useful starting point for the non-relativistic many-body theory based on the quanitzed Schrödinger field.

## 2. THE SCHRÖDINGER FIELD

Let us denote the Schrödinger field by $\psi(\vec{x},t)$, and assume the Lagrangian density

$$\mathcal{L}_0(x,t) = i\hbar\psi^\dagger(\vec{x},t)\frac{\partial}{\partial t}\psi(\vec{x},t) - \frac{\hbar^2}{2m}\vec{\nabla}\psi^\dagger(\vec{x},t)\cdot\vec{\nabla}\psi(\vec{x},t)$$

$$- \frac{1}{2}\int d^3x'\psi^\dagger(\vec{x},t)\psi^\dagger(\vec{x}',t)V(|\vec{x}-\vec{x}'|)\psi(\vec{x}',t)\psi(\vec{x},t)$$

$$- V^{ext}(\vec{x})\psi^\dagger(\vec{x},t)\psi(\vec{x},t) \tag{2.1}$$

where $V(|\vec{x}-\vec{x}'|)$ is the potential between two particles at $\vec{x}$

and $\vec{x}'$ and $V^{ext}(\vec{x})$ the potential due to external sources*). To derive the field equation, we impose on (2.1) the Hamiltonian principle and obtain

$$\partial_t \frac{\partial \mathcal{L}_o}{\partial \partial_t \psi^\dagger(\vec{x},t)} + \vec{\nabla}\cdot \frac{\partial \mathcal{L}_o}{\partial \vec{\nabla}\psi^\dagger(\vec{x},t)} - \frac{\partial \mathcal{L}_o}{\partial \psi^\dagger(\vec{x},t)}$$

$$= \{-\frac{\hbar^2}{2m}\nabla^2 - i\hbar\frac{\partial}{\partial t}\}\psi(\vec{x},t) + v_o(\vec{x},t)\psi(\vec{x},t)$$

$$+ V^{ext}(\vec{x})\psi(\vec{x},t) = 0 , \qquad (2.2a)$$

*) Since we are dealing with the quantized Schrödinger field, the order of product of two fields cannot be interchanged arbitrarily.

$$\partial_t \frac{\partial \mathcal{L}_o}{\partial \partial_t \psi(\vec{x},t)} + \vec{\nabla}\cdot \frac{\partial \mathcal{L}_o}{\partial \vec{\nabla}\psi(\vec{x},t)} - \frac{\partial \mathcal{L}_o}{\partial \psi(\vec{x},t)}$$

$$= \{i\hbar\frac{\partial}{\partial t} - \frac{\hbar^2}{2m}\nabla^2\}\psi^\dagger(\vec{x},t) + \psi^\dagger(\vec{x},t)v_o(\vec{x},t)$$

$$+ \psi^\dagger(\vec{x},t)V^{ext}(\vec{x}) = 0 \qquad (2.2b)$$

where $v_o(\vec{x},t)$ stands for

$$v_o(\vec{x},t) = \int d^3x' \; V(|\vec{x}-\vec{x}'|)\psi^\dagger(\vec{x}',t)\psi(\vec{x}',t) . \qquad (2.3)$$

The canonical momentum of the field operator $\psi(\vec{x},t)$ is defined as

$$\Pi(\vec{x},t) = \frac{\partial \mathcal{L}_o}{\partial \partial_t \psi^\dagger(\vec{x},t)} = i\hbar\psi^\dagger(\vec{x},t) . \qquad (2.4)$$

Hence, the Hamiltonian density is

$$\mathcal{H}_o(\vec{x}.t) \equiv \Pi(\vec{x},t)\,\frac{\partial}{\partial t}\,\psi(\vec{x},t) - \mathcal{L}_o(\vec{x},t)$$

$$= \frac{\hbar^2}{2m}\,\vec{\nabla}\psi^\dagger(\vec{x},t)\cdot\vec{\nabla}\psi(\vec{x},t)$$

$$+ \frac{1}{2}\,\psi^\dagger(\vec{x},t)v_o(\vec{x},t)\psi(\vec{x},t)$$

$$+ V^{ext}(\vec{x})\psi^\dagger(\vec{x},t)\psi(\vec{x},t) \ . \tag{2.5}$$

Depending on the statistics obeyed by the field, we set the commutator

$$[\psi(\vec{x},t),\psi^\dagger(\vec{x}',t)] = \delta(\vec{x}-\vec{x}')$$

(for Bose statistics) (2.6a)

or the anticommutator

$$\{\psi(\vec{x},t),\psi^\dagger(\vec{x}',t)\} = \delta(\vec{x}-\vec{x}')$$

(for Fermi statistics). (2.6b)

The following formulas do not depend on the type of statistics, unless otherwise stated.

Let us define useful physical quantities.

(i) The number density and its flow is defined as a generating current of the phase transformation, and they are

$$J_o(\vec{x},t) = \psi^\dagger(\vec{x},t)\psi(\vec{x},t) \tag{2.7a}$$

$$J_i(\vec{x},t) = -i\,\frac{\hbar}{2m}\,\{\psi^\dagger(\vec{x},t)\,\frac{\partial}{\partial x_i}\,\psi(\vec{x},t)$$

$$- \frac{\partial}{\partial x_i}\,\psi^\dagger(\vec{x},t)\cdot\psi(\vec{x},t)\} \ . \tag{2.7b}$$

It should be noted that the quantities (2.7) do not contain the potentials $V(|\vec{x}-\vec{x}'|)$ and $V^{ext}(\vec{x})$ explicitly. It is easy to prove by the help of the field equation (2.2) that the continuity equation

$$\frac{\partial}{\partial t} J_o(\vec{x},t) + \vec{\nabla}\cdot\vec{J}(\vec{x},t) = 0 \qquad (2.8)$$

holds true. This means that the field equation (2.2) is valid when the number of the particles involved in the system does not change.

(ii) The so-called <u>canonical energy-momentum</u> can be defined as a generating current of the space-time translations. They are

$$\begin{aligned} T_{oi}(\vec{x},t) &= -\frac{\partial \mathcal{L}_o}{\partial \partial_t \psi}\,\partial_i \psi - \partial_i \psi^\dagger \frac{\partial \mathcal{L}_o}{\partial \partial_t \psi^\dagger} \\ &= -i\hbar \psi^\dagger(\vec{x},t)\,\partial_i \psi(\vec{x},t) \ , \end{aligned} \qquad (2.9a)$$

$$\begin{aligned} T_{ji}(\vec{x},t) &= -\frac{\partial \mathcal{L}_o}{\partial \partial_j \psi}\,\partial_i \psi - \partial_i \psi^\dagger \frac{\partial \mathcal{L}_o}{\partial \partial_j \psi^\dagger} + \delta_{ij}\mathcal{L}_o \\ &= \frac{\hbar^2}{2m}\left\{\partial_j \psi^\dagger(\vec{x},t)\,\partial_i \psi(\vec{x},t) + \partial_i \psi^\dagger(\vec{x},t)\,\partial_j \psi(\vec{x},t)\right\} \\ &\quad + \delta_{ij}\mathcal{L}_o(\vec{x},t) \ , \end{aligned} \qquad (2.9b)$$

$$T_{oo}(\vec{x},t) = \frac{\partial \mathcal{L}_o}{\partial \partial_t \psi} \partial_t \psi + \partial_t \psi^\dagger \frac{\partial \mathcal{L}_o}{\partial \partial_t \psi^\dagger} - \mathcal{L}_o$$

$$= \frac{\hbar^2}{2m} \vec{\nabla}\psi^\dagger(\vec{x},t)\cdot\vec{\nabla}\psi(\vec{x},t)$$

$$+ \frac{1}{2} \psi^\dagger(\vec{x},t)v_o(\vec{x},t)\psi(\vec{x},t)$$

$$+ V^{ext}(\vec{x})\psi^\dagger(\vec{x},t)\psi(\vec{x},t)$$

$$= \mathcal{H}_o(\vec{x},t) \qquad (2.10a)$$

$$T_{jo}(\vec{x},t) = \frac{\partial \mathcal{L}_o}{\partial \partial_j \psi} \partial_t \psi + \partial_t \psi^\dagger \frac{\partial \mathcal{L}_o}{\partial \partial_j \psi^\dagger}$$

$$= -\frac{\hbar^2}{2m} \{\partial_j\psi^\dagger(\vec{x},t)\partial_t\psi(\vec{x},t)+\partial_t\psi^\dagger(\vec{x},t)\partial_j\psi(\vec{x},t)\}$$

$$= \frac{i}{\hbar}\frac{\hbar^2}{2m} \{\partial_j\psi^\dagger(\vec{x},t)(-\frac{\hbar^2}{2m}\nabla^2+v_o(\vec{x},t)+V^{ext}(\vec{x}))\psi(\vec{x},t)\}$$

$$-\frac{i}{\hbar}\frac{\hbar^2}{2m} \{(-\frac{\hbar^2}{2m}\nabla^2\psi^\dagger(\vec{x},t)+\psi^\dagger(\vec{x},t)v_o(\vec{x},t)$$

$$+ \psi^\dagger(\vec{x},t)V^{ext}(\vec{x}))\partial_j\psi(\vec{x},t)\} \qquad (2.10b)$$

where

$$\partial_i = \frac{\partial}{\partial x_i}$$

$$\partial_t = \frac{\partial}{\partial t}$$

and also the field equation (2.2) has been used to eliminate the time derivatives (a requirement in canonical theory). The physical meanings of these quantities are

$T_{oi}(\vec{x},t)$: The momentum density along i-th axis.

$T_{ji}(\vec{x},t)$: The j-th component of flow of momentum density along i-th axis.

$T_{oo}(\vec{x},t)$: The energy density.

$T_{jo}(\vec{x},t)$: The j-th component of energy flow.

It is expected that the homogeneity of the space and time leads to the usual continuity equations. However, in the presence of the potential interaction, the momentum density is not conserved locally, though the total momentum is conserved in the absence of external potential. To see this, we calculate*)

$$\partial_t T_{oi}(\vec{x},t) + \partial_j T_{ji}(\vec{x},t) = - \psi^\dagger(\vec{x},t)\partial_i v_o(\vec{x},t)\psi(\vec{x},t)$$

$$- \partial_i V^{ext}(\vec{x})\psi^\dagger(\vec{x},t)\psi(\vec{x},t) . \qquad (2.11)$$

*) Summation convention is used hereafter, i.e. sum over repeated indices.

To obtain the last expression, the field equation (2.2) has to be used. Eq. (2.11) shows clearly, the momentum density is not conserved locally due to the presence of the potential even in the absence of $V^{ext}(\vec{x})$. But, let us integrate both sides of eq. (2.11) to see whether the total momentum is globally conserved. Denoting the total momentum by $P_i$, we have

$$P_i = \int d^3x \; T_{oi}(\vec{x},t) \; . \qquad (2.12)$$

Then, eq. (2.11) implies

$$\frac{d}{dt} P_i = -\int d^3x \, \partial_j T_{ji}(\vec{x},t)$$

$$- \int d^3x \, d^3x' \psi^\dagger(\vec{x},t)\psi^\dagger(\vec{x}',t)\partial_i V(|\vec{x}-\vec{x}'|)$$

$$\times \psi(\vec{x}',t)\psi(\vec{x},t)$$

$$- \int d^3x \, \partial_i V^{ext}(\vec{x})\psi^\dagger(\vec{x},t)\psi(\vec{x},t) \ . \qquad (2.13)$$

The first term can be converted to a surface integral by the help of the Gauss theorem. The second term vanishes as follows: First, interchange $\vec{x}$ and $\vec{x}'$ and use the equal time commutator (or anticommutator)

$$\psi(\vec{x},t)\psi(\vec{x}',t) \mp \psi(\vec{x}',t)\psi(\vec{x},t) = 0 \qquad (2.14a)$$

$$\psi^\dagger(\vec{x},t)\psi^\dagger(\vec{x}',t) \mp \psi^\dagger(\vec{x}',t)\psi^\dagger(\vec{x},t) = 0 \qquad (2.14b)$$

and

$$\partial_i V(|\vec{x}-\vec{x}'|) = -\partial_i' V(|(\vec{x}-\vec{x}'|) \ . \qquad (2.15)$$

Due to the negative sign in eq. (2.15), the result is negative of the original expression. Hence it is zero. We can then state that <u>the total momentum (2.12) vanishes provided the surface integral can be ignored and</u> $\partial_i V^{ext}=0$. Note that the relation (2.15) is nothing but Newton's third law. The symmetry of $T_{ij}(\vec{x},t)$ is a consequence of the spherical symmetry and hence the angular momentum conservation law.

The proof of energy conservation is straightforward, as long as the potentials are independent of time, since continuity equation

$$\partial_t T_{oo}(\vec{x},t) + \partial_j T_{jo}(\vec{x},t) = 0 \tag{2.16}$$

can easily be established by virtue of the field equation (2.2), i.e.,

$$\frac{d}{dt} H_o = 0 \tag{2.17}$$

where

$$H_o = \int d^3x \; T_{oo}(\vec{x},t) \quad . \tag{2.18}$$

Here again, the surface integral has been neglected.

(iii) The <u>velocity density</u>

$$J_i(\vec{x},t) = -i \frac{\hbar^2}{2m} \{\psi^\dagger(\vec{x},t)\partial_i\psi(\vec{x},t) - \partial_i\psi^\dagger(\vec{x},t)\psi(\vec{x},t)\}$$

$$\equiv -i \frac{\hbar^2}{2m} \psi^\dagger(\vec{x} \;,t)(\partial_i - \overleftarrow{\partial}_i)\psi(\vec{x},t) \tag{2.19}$$

which appeared in (2.7b), and its flow

$$J_{ji}(\vec{x},t) = -(\frac{\hbar}{2m})^2 \psi^\dagger(\vec{x},t)(\partial_j - \overleftarrow{\partial}_j)(\partial_i - \overleftarrow{\partial}_i)\psi(\vec{x},t) \tag{2.20}$$

satisfy the balance equation

$$\partial_t J_i(\vec{x},t) - \partial_j J_{ji}(\vec{x},t) = -\frac{1}{m} \psi^\dagger(\vec{x},t)\partial_i v_o(\vec{x},t)\psi(\vec{x},t)$$

$$- \frac{1}{m} \partial_i V^{ext}(\vec{x}) J_o(\vec{x},t) \quad . \tag{2.21}$$

The first term in the right-hand-side vanishes upon integration as was demonstrated in the preceding example. (Note however, the presence of this term makes hydrodynamical limit a little tricky.) Hence the acceleration of the total system is equal to the sum of forces acting locally (Newton's second law), i.e.

$$\frac{d}{dt}\int d^3x\ J_i(\vec{x},t) = -\frac{1}{m}\int d^3x\ \partial_i V^{ext}(\vec{x})J_o(\vec{x},t)\ . \qquad (2.22)$$

This property is related to the invariance of our theory under the Galilei transformation which we shall elaborate on in the next section. Some of the commutators among (2.7), (2.9a) and (2.10a) are listed in Appendix A.

## 3. THE GALILEI TRANSFORMATION*)

Under the Galilei transformation

$$\vec{x} \to \vec{x}\,' = \vec{x} - \vec{u}t \qquad (3.1a)$$

$$t \to t' = t \qquad (3.1b)$$

the Schrödinger field undergoes the transformation

$$\psi(\vec{x},t) \to \psi'(\vec{x}\,',t') = S_G(\vec{x},t)\psi(\vec{x},t) \qquad (3.2a)$$

$$\psi^\dagger(\vec{x},t) \to \psi^{\dagger\prime}(\vec{x}\,',\vec{t}\,') = S_G^*(\vec{x},t)\psi^\dagger(\vec{x},t) \qquad (3.2b)$$

with

$$S_G(\vec{x},t) = \exp\{-i(m\vec{u}\cdot\vec{x} - \tfrac{1}{2}\,mu^2t)/\hbar\} \qquad (3.3)$$

where $\vec{u}$ is the velocity of the primed system relative to the unprimed system. The derivative transform as

$$\partial_i \to \partial_i' = \frac{\partial x_j}{\partial x_i'}\,\partial_j + \frac{\partial t}{\partial x_i'}\,\partial_t = \partial_i\ , \qquad (3.4a)$$

$$\partial_t \to \partial_t' = \frac{\partial x_j}{\partial t'}\,\partial_j + \frac{\partial t}{\partial t'}\,\partial t = u_j\partial_j + \partial_t\ . \qquad (3.4b)$$

---

*) We put $V^{ext}(\vec{x})=0$ in this section.

Hence, we obtain

$$\partial_i \psi(\vec{x},t) \rightarrow \partial_i' \psi'(\vec{x}',t') = S_G(\vec{x},t)(\partial_i - imu_i/\hbar)\psi(\vec{x},t) \quad (3.5a)$$

$$\partial_i \ \psi^\dagger(\vec{x},t) \rightarrow \partial_i' \psi^{\dagger '}(\vec{x}',\vec{t}') = S_G^*(\vec{x},t)(\partial_i + imu_i/\hbar)\psi^\dagger(\vec{x},t) \quad (3.5b)$$

$$\begin{aligned}\partial_t \psi(\vec{x},t) &\rightarrow \partial_{t'} \psi'(\vec{x}',t') \\ &= S_G(\vec{x},t)(\partial_t + u_j\partial_j + i\,\frac{1}{2}\,mu^2/\hbar - imu^2/\hbar)\psi(\vec{x},t) \\ &= S_G(\vec{x},t)(\partial_t + u_j\partial_j - \frac{1}{2}\,imu^2/\hbar)\psi(\vec{x},t). \end{aligned} \quad (3.5c)$$

On substituting eq. (3.5), we can easily prove that the Lagrangian density (2.1) is invariant under the Galilei transformation (3.1) and (3.2), i.e.,

$$\mathcal{L}_o(\vec{x},t) \rightarrow \mathcal{L}_o'(\vec{x}',t') = \mathcal{L}_o(\vec{x},t) \ . \quad (3.6)$$

According to the Noether theorem, there is a conserved quantity associatd with the Galilei transformation*). Using the standard Noether method, we define

---

*) It is often seen in the literature that the Schrödinger field equation (2.2) is shown to be invariant under the Galilei transformation. However, we emphasize here that the invariance of the field equation is not sufficient to show that there is a conserved quantity. See for example reference 4).

---

$$\begin{aligned} N_{oi}(\vec{x},t) &= -\{mx_i\psi^\dagger(\vec{x},t)\psi(\vec{x},t) + i\hbar t\psi^\dagger(\vec{x},t)\partial_i\psi(\vec{x},t)\} \\ &= -\{mx_i J_o(\vec{x},t) - tT_{oi}(\vec{x},t)\} \end{aligned} \quad (3.7a)$$

$$N_{ji}(\vec{x},t) = imx_i \frac{\hbar}{2m} \psi^\dagger(\vec{x},t)(\partial_j - \overleftarrow{\partial}_j)\psi(\vec{x},t)$$

$$+ t\{\frac{\hbar^2}{2m}(\partial_i\psi^\dagger(\vec{x},t)\partial_j\psi(\vec{x},t) + \partial_j\psi^\dagger(\vec{x},t)\partial_i\psi(\vec{x},t))$$

$$+ \delta_{ij}\mathcal{L}_o(\vec{x},t)\}$$

$$= -\{mx_i J_j(\vec{x},t) - tT_{ji}(\vec{x},t)\} . \qquad (3.7b)$$

The balance equation is

$$\partial_t N_{oi}(\vec{x},t) + \partial_j N_{ji}(\vec{x},t)$$

$$= -\{mx_i(\partial_t J_o(\vec{x},t) + \partial_j J_j(\vec{x},t)$$

$$- t(\partial_t T_{oi}(\vec{x},t) + \partial_j T_{ji}(\vec{x},t))$$

$$- T_{oi}(\vec{x},t) + mJ_i(\vec{x},t)\} . \qquad (3.8)$$

On account of the relations (2.8) and (2.11), and the forms (2.7b) and (2.9a), we have the final expression of the balance relation

$$\partial_t N_{oi}(\vec{x},t) + \partial_j N_{ji}(\vec{x},t)$$

$$= -\frac{t}{2} \psi^\dagger(\vec{x},t)\partial_i v_o(\vec{x},t)\psi(\vec{x},t)$$

$$- \frac{i}{2}\hbar\partial_i J_o(\vec{x},t) . \qquad (3.9)$$

The local conservation of the Galilei current (3.7) is not satisfied again due to the presence of the potential.

The conservation of the Galilei generator

$$G_i \equiv \int d^3x\, N_{oi}(\vec{x},t) \qquad (3.10)$$

can be proved in the absence of external potential in a similar fashion with the case of the momentum using Newton's third law. We note that

$$-i\hbar\, \delta_i^L \psi(\vec{x},t) = [\psi(\vec{x},t), G_i] \tag{3.11a}$$

$$-i\hbar \delta_i^L \psi^\dagger(\vec{x},t) = [\psi^\dagger(\vec{x},t), G_i] \tag{3.11b}$$

on account of the commutators (2.6), where

$$\delta_i^L \psi(\vec{x},t) \equiv (-i \frac{m}{\hbar} x_i + t\partial_i)\psi(\vec{x},t) \ , \tag{3.12a}$$

$$\delta_i^L \psi^\dagger(\vec{x},t) \equiv (i \frac{m}{\hbar} x_i + t\partial_i)\psi^\dagger(\vec{x},t) \ . \tag{3.12b}$$

As a final remark, we calculate the commutator of the $G_i$ and the total Hamiltonian (2.18). Since the $G_i$ depends explicitly on t, we have

$$\frac{dG_i}{dt} = \frac{\partial G_i}{\partial t} + \frac{1}{i\hbar} [G_i, H_o] \tag{3.13}$$

where in the first term in the right-hand-side the time derivative is taken only with respect to the explicit t, namely,

$$\frac{\partial G_i}{\partial t} = P_i \ . \tag{3.14}$$

If we calculate the commutators of $G_i$ with $J_o(\vec{x},t)$, $J_i(\vec{x},t)$, $T_{io}(\vec{x},t)$, then we obtain

$$[J_o(\vec{x},t) \ , G_j] = -it\hbar\partial_j J_o(\vec{x},t) \tag{3.15}$$

$$[J_i(\vec{x},t),G_j)] = i\hbar\{\delta_{ij}J_o(\vec{x},t)-t\partial_j J_i(\vec{x},t)\} \qquad (3.16)$$

$$[T_{io}(\vec{x},t),G_j] = i\hbar\{\delta_{ij}T_{oo}(\vec{x},t)$$

$$-T_{ij}(\vec{x},t)-t\partial_i T_{jo}(\vec{x},t)\} \ . \qquad (3.17)$$

These relations are useful in the investigation of the spontaneous breakdown of the Galilei invariance[5)*)]

---

*) Before the theory of relativity was proposed by Einstein, the invariances of the Newton mechanics and the Maxwell theory were not compatible: The former is invariant under the Galilei transformation whereas the latter is invariant under the Lorentz transformation. As is well-known, Einstein dissolved this dilemma by rewriting the mechanics in such a way that it is invariant with respect to the Lorentz transformation. Could it be a possible scenario that if someone tried to eliminate the light velocity from the Maxwell equation and rewrite the vector field equation in a Galilei invariant form instead, he would have discovered the Schrödinger equation of the type (4.8), presented in the next section?

---

## 4. AN ALTERNATIVE FORM OF THE SCHRÖDINGER EQUATION

We have been tacitly assuming that the Schrödinger field $\psi(\vec{x},t)$ is of a single component. In order to consider a nonrelativistic particle with spin 1/2, we have to introduce a field with 2 component $\psi_\alpha(\vec{x},t)$ ($\alpha$=1,2), which satisfies the Schrödinger equation

$$i\hbar\partial_t\psi_\alpha(\vec{x},t) = -\frac{\hbar^2}{2m}\nabla^2\psi_\alpha(\vec{x},t) \qquad (\alpha=1,2) \ . \qquad (4.1)$$

Equation (4.1) can be decomposed as

$$i\hbar\partial_t\psi(\vec{x},t) = -i\hbar\,\vec{\sigma}\cdot\vec{\nabla}\,\chi(\vec{x},t) \qquad (4.2a)$$

$$2m\ \chi(x,t) = -i\hbar\ \vec{\sigma}\cdot\vec{\nabla}\ \psi(\vec{x},t)\ . \tag{4.2b}$$

Here, we have used a matrix notation to avoid spin indices of two-component fields $\psi$ and $\chi$. The $\vec{\sigma}$ are the well-known Pauli spin matrices, i.e.

$$\sigma_1 = \begin{pmatrix} 0 & 1 \\ 1 & 0 \end{pmatrix} \qquad \sigma_2 = \begin{pmatrix} 0 & -i \\ i & 0 \end{pmatrix} \qquad \sigma_3 = \begin{pmatrix} 1 & 0 \\ 0 & -1 \end{pmatrix} \tag{4.3}$$

which satisfy

$$[\sigma_i, \sigma_j] = 2i\varepsilon_{ijk}\sigma_k \tag{4.4a}$$

$$\{\sigma_i, \sigma_j\} = 2\delta_{ij}\ . \tag{4.4b}$$

Due to eq. (4.4), it is easy to prove

$$(\vec{\sigma}\cdot\vec{\nabla})^2 = I\nabla^2\ . \tag{4.5}$$

Eliminating $\chi$ from (4.2) and using the identity (4.5), we can easily show that eq. (4.2) reduces to eq. (4.1).

The Lagrangian density leading to eq. (4.2) can readily be constructed:

$$\begin{aligned} \mathcal{L}_0(\vec{x},t) = {} & i\hbar\ \psi^\dagger(\vec{x},t)\partial_t\psi(\vec{x},t) + 2m\ \chi^\dagger(\vec{x},t)\chi(\vec{x},t) \\ & + i\hbar\ \psi^\dagger(\vec{x},t)\vec{\sigma}\cdot\vec{\nabla}\ \chi(\vec{x},t) \\ & - i\hbar\ \vec{\nabla}\chi^\dagger(\vec{x},t)\cdot\vec{\sigma}\psi(\vec{x},t)\ . \end{aligned} \tag{4.6}$$

We leave it as an exercise to derive eq. (4.2) from (4.6). In this regard, note that in eq. (4.6), we simplified the expression by omitting indices of the fields, i.e.

$$\psi^\dagger\psi \equiv \psi^\dagger_\alpha\psi_\alpha \tag{4.7a}$$

$$\chi^\dagger\chi \equiv \chi^\dagger_\alpha\chi_\alpha\ , \quad \text{etc.} \tag{4.7b}$$

The physical quantities such as the number density, the energy density, etc. can be defined. Notice however that eq. (4.2b) is merely defining the field $\chi_\alpha(\vec{x},t)$ and containing no time derivative. This fact makes the canonical formalism slightly more complicated, and hence the above formalism based on the equation of the type (4.2) seems to be quite useless from the practical view point. But, that it is not so will be seen when we introduce the electromagnetic interaction via the minimal coupling. Before returning to the realistic problem of electromagnetic interaction, we point out that eq. (4.2) is not the only alternative form of the Schrödinger field. For example, consider a vector field $\psi_i(\vec{x},t)$ $(i=1,2,3)$ satisfying

$$i\hbar\,\partial_t\psi_i(\vec{x},t) = -\frac{\hbar^2}{2m_T}\left(\nabla^2\delta_{ik}-\partial_i\partial_k\right)\psi_k(\vec{x},t) - \frac{\hbar^2}{2m_L}\,\partial_i\partial_k\psi_k(\vec{x},t)\ . \tag{4.8}$$

It is not difficult to deduce from eq. (4.8) that

$$i\hbar\partial_t\phi(\vec{x},t) = -\frac{\hbar^2}{2m_L}\nabla^2\phi(\vec{x},t) \tag{4.9}$$

and

$$i\hbar\partial_t\phi_i(\vec{x},t) = -\frac{\hbar^2}{2m_T}\nabla^2\phi_i(\vec{x},t) \tag{4.10}$$

with

$$\phi(\vec{x},t) \equiv \partial_k \psi_k(\vec{x},t) \tag{4.11a}$$

$$\phi_i(\vec{x},t) \equiv \varepsilon_{ijk}\partial_j \psi_k(\vec{x},t) \tag{4.11b}$$

which are longitudinal and transverse fields, having the masses $m_L$ and $m_T$, respectively. This field $\psi_i(\vec{x},t)$ carries spin 1. Incidentally, eq. (4.8) can also be reduced to equations with only first order derivatives. They are, as is easily verified,

$$i\hbar\, \partial_t \psi_i(\vec{x},t) = -\frac{\hbar^2}{2m_T}\varepsilon_{ijk}\partial_j \phi_k(\vec{x},t) - \frac{\hbar^2}{2m_L}\partial_i \phi(\vec{x},t) \tag{4.12a}$$

$$\phi(\vec{x},t) = \partial_k \psi_k(\vec{x},t) \tag{4.12b}$$

$$\phi_i(\vec{x},t) = \varepsilon_{ijk}\partial_j \psi_k(\vec{x},t) \; . \tag{4.12c}$$

An extension to the fields with integer higher spin is quite straightforward, if necessary. There are a number of suggestions to extend the formalism to fields with half-odd integer spin. We shall only refer the readers to reference 6.

## 5. ELECTROMAGNETIC INTERACTION

If the Schrödinger field $\psi(\vec{x},t)$ is charged, it interacts with the electromagnetic field. To introduce electromagnetic interaction, we usually adopt the principle of minimal electromagnetic coupling, i.e., replace the momentum $\vec{p}$ by $\vec{p}-e\vec{A}/c$ and the energy H by $H-eA_o$ where $\vec{A}$ and $A_o$ are vector and scalar potentials, respectively. The field theoretical version of the minimal electromagnetic coupling is

$$\vec{\nabla} \to \vec{\nabla} - i \frac{e}{\hbar c} \vec{A}(\vec{x},t) \tag{5.1a}$$

$$i\hbar\partial_t \to i\hbar\partial_t - eA_o(\vec{x},t) \ . \tag{5.1b}$$

Hence, the Schrödinger equation in the presence of the electromagnetic field is

$$i\hbar \ \partial_t\psi(\vec{x},t) = -\frac{\hbar^2}{2m}\left(\vec{\nabla}-i\frac{e}{\hbar c}\vec{A}(\vec{x},t)\right)^2\psi(\vec{x},t) + eA_o(\vec{x},t)\psi(\vec{x},t) \ . \tag{5.2}$$

This equation is invariant under the gauge transformation

$$\vec{A} \to \vec{A} + \vec{\nabla}\lambda(\vec{x},t) \tag{5.3a}$$

$$A_o \to A_o - \frac{1}{c}\partial_t\lambda(\vec{x},t) \tag{5.3b}$$

$$\psi(\vec{x},t) \to \exp\{e\lambda(\vec{x},t)/\hbar c\}\psi(\vec{x},t) \ . \tag{5.3c}$$

But, from the gauge principle, we can add arbitrarily to eq. (5.2) a term such as

$$-g\mu_e \frac{\vec{\sigma}}{2} \cdot \vec{H}(\vec{x},t)\psi(\vec{x},t) \tag{5.4}$$

with the Bohr magneton

$$\mu_e \equiv \frac{e\hbar}{2mc} \ , \tag{5.5}$$

where $\vec{H}$ is the magnetic field and $\vec{\sigma}$ the Pauli spin matrices. The factor g is arbitrary from the gauge principle only. But, the analysis of the fine structure of atomic spectra shows that the g-factor must be 2 for the electron. There seems no principle in the Schrödinger theory that gives g=2. If we introduce the minimal coupling (5.1) into the relativistic Dirac electron equation, we obtain precisely g=2! This has been considered a triumphant feature of the relativistic theory.

However, let us look at the electron equation of the 1st order form (4.2) and apply the minimal principle (5.1):

$$i\hbar\, \partial_t \psi(\vec{x},t) = -i\hbar\, \vec{\sigma}\cdot\left(\vec{\nabla} - i\,\frac{e}{\hbar c}\,\vec{A}(\vec{x},t)\right)\chi(\vec{x},t) + eA_o(\vec{x},t)\psi(\vec{x},t) \qquad (5.6a)$$

$$2m\chi(\vec{x},t) = -i\hbar\vec{\sigma}\cdot\left(\vec{\nabla} - i\,\frac{e}{\hbar c}\,\vec{A}(\vec{x},t)\right)\psi(\vec{x},t) \;. \qquad (5.6b)$$

Now, to combine these two equations we carry out the following manipulation:

$$\varepsilon_{ijk}\left(\partial_i - i\,\frac{e}{\hbar c}\,A_i(\vec{x},t)\right)\left(\partial_j - i\,\frac{e}{\hbar c}\,A_j(\vec{x},t)\right) = -i\,\frac{e}{\hbar c}\,H_k(\vec{x},t) \qquad (5.7)$$

and, using eqs. (4.4) and (5.7),

$$\begin{aligned}\{\vec{\sigma}\cdot(\vec{\nabla} - i\,\tfrac{e}{\hbar c}\,\vec{A}(\vec{x},t))\}^2 &= \sigma_i\sigma_j\left(\partial_i - i\,\tfrac{e}{\hbar c}\,A_i(\vec{x},t)\right)\left(\partial_j - i\,\tfrac{e}{\hbar c}\,A_j(\vec{x},t)\right) \\ &= (\delta_{ij} + i\varepsilon_{ijk}\sigma_k)\left(\partial_i - i\,\tfrac{e}{\hbar c}\,A_i(\vec{x},t)\right)\left(\partial_j - i\,\tfrac{e}{\hbar c}\,A_j(\vec{x},t)\right) \\ &= \left(\vec{\nabla} - i\,\tfrac{e}{\hbar c}\,\vec{A}(\vec{x},t)\right)^2 + \tfrac{e}{\hbar c}\,\vec{\sigma}\cdot\vec{H}(\vec{x},t) \;. \end{aligned} \qquad (5.8)$$

Using eq. (5.8), we eliminate $\chi(\vec{x},t)$ from eq. (5.6) to obtain

$$i\hbar\partial_t\psi(\vec{x},t) = -\frac{\hbar^2}{2m}\left(\vec{\nabla}-i\frac{e}{\hbar c}\vec{A}(\vec{x},t)\right)^2\psi(\vec{x},t) + eA_o(\vec{x},t)\psi(\vec{x},t) - \frac{e\hbar}{2mc}\vec{\sigma}\cdot\vec{H}(\vec{x},t)\psi(\vec{x},t). \qquad (5.9)$$

The comparison of eq. (5.4) with the last term of eq. (5.9) immediately gives

$$g = 2 . \qquad (5.10)$$

So, even in non-relativistic theory, if we start from the 1st order equation and employ the minimal principle, we obtain the required relation (5.10). This important fact was pointed out by Levy-Leblond in 1967[1)], but unfortunately very little attention has been drawn.

Finally, let us define the charge and current densities in the presence of electromagnetic field. For this purpose, let us apply the minimal principle (5.1) to the Lagrangian (4.6), to obtain

$$\mathcal{L}(\vec{x},t) = \mathcal{L}_o(\vec{x},t) + \mathcal{L}_{int}(\vec{x},t) \qquad (5.11)$$

with

$$\mathcal{L}_{int}(\vec{x},t) = -eA_o(\vec{x},t)\psi^\dagger(\vec{x},t)\psi(\vec{x},t) - \frac{e}{c}\psi^\dagger(\vec{x},t)\vec{\sigma}\chi(\vec{x},t)\cdot\vec{A}(\vec{x},t) - \frac{e}{c}\chi^\dagger(\vec{x},t)\vec{\sigma}\psi(\vec{x},t)\cdot\vec{A}(\vec{x},t) . \qquad (5.12)$$

In this case, eq. (5.6) follows from the Hamiltonian

principle. To define the charge density, we calculate

$$j_o(x,t) \equiv \frac{e}{i\hbar}\left(\frac{\partial\mathcal{L}}{\partial\partial_t\psi}\psi - \psi^\dagger\frac{\partial\mathcal{L}}{\partial\partial_t\psi^\dagger} + \frac{\partial\mathcal{L}}{\partial\partial_t\chi}\chi - \chi^\dagger\frac{\partial\mathcal{L}}{\partial\partial_t\chi^\dagger}\right)$$

$$= e\psi^\dagger(\vec{x},t)\psi(\vec{x},t) \quad , \tag{5.13a}$$

$$j_i(x,t) \equiv \frac{e}{i\hbar}\left(\frac{\partial\mathcal{L}}{\partial\partial_i\psi}\psi - \psi^\dagger\frac{\partial\mathcal{L}}{\partial\partial_i\psi^\dagger} + \frac{\partial\mathcal{L}}{\partial\partial_i\chi}\chi - \chi^\dagger\frac{\partial\mathcal{L}}{\partial\partial_i\chi^\dagger}\right)$$

$$= e\{\psi^\dagger(\vec{x},t)\sigma_i\chi(\vec{x},t)+\chi^\dagger(\vec{x},t)\sigma_i\psi(\vec{x},t)\} \quad . \tag{5.13b}$$

Using (5.6b), we eliminate $\chi$ and $\chi^\dagger$ from eq. (5.13b). Then

$$\begin{aligned} j_i(x,t) &= -i\,\frac{e\hbar}{2m}\,\{\psi^\dagger(\vec{x},t)\sigma_i\vec{\sigma}\cdot(\vec{\nabla}-i\,\frac{e}{\hbar c}\,\vec{A}(\vec{x},t))\psi(\vec{x},t) \\ &\quad - (\vec{\nabla}+i\,\frac{e}{\hbar c}\,\vec{A}(\vec{x},t))\psi^\dagger(\vec{x},t)\cdot\vec{\sigma}\sigma_i\psi(\vec{x},t)\} \\ &= -i\,\frac{e\hbar}{2m}\,\{\psi^\dagger(\vec{x},t)(\partial_i-i\,\frac{e}{\hbar c}\,A_i(\vec{x},t))\psi(\vec{x},t) \\ &\quad - (\partial_i+i\,\frac{e}{\hbar c}\,A_i(\vec{x},t))\psi^\dagger(\vec{x},t)\cdot\psi(\vec{x},t)\} \\ &\quad + \frac{e\hbar}{2m}\,\varepsilon_{ijk}\partial_j(\psi^\dagger(\vec{x},t)\sigma_k\psi(\vec{x},t)) \end{aligned} \tag{5.14}$$

or in more familiar notation

$$\begin{aligned} \vec{j}(\vec{x},t) &= -i\,\frac{e\hbar}{2m}\,\{\psi^\dagger(\vec{x},t)(\vec{\nabla}-i\,\frac{e}{\hbar c}\,\vec{A}(\vec{x},t))\psi(\vec{x},t) \\ &\quad - (\vec{\nabla}+i\,\frac{e}{\hbar c}\,\vec{A}(\vec{x},t))\psi^\dagger(\vec{x},t)\cdot\psi(\vec{x},t)\} \\ &\quad + \frac{e\hbar}{2m}\,\vec{\nabla}\times(\psi^\dagger(\vec{x},t)\vec{\sigma}\psi(\vec{x},t)) \quad . \end{aligned} \tag{5.15}$$

The last term is the contribution to the current from the magnetic moment. It can easily be verified that the charge and the current above satisfy the continuity equation.

It is an interesting exercise to derive the equation corresponding to Lorentz's force law, i.e.

$$\frac{d}{dt}\int d^3x j_i(\vec{x},t) = \frac{e}{m}\int d^3x\{j_o(\vec{x},t)E_i(\vec{x},t)+\frac{1}{c}(\vec{j}(\vec{x},t)\times\vec{H}(\vec{x},t))\}$$

$$-\frac{1}{m}\int d^3x\partial_i V^{ext}(\vec{x})j_o(\vec{x},t) \qquad (5.16)$$

which holds even when the internal potential $V(|\vec{x}-\vec{x}'|)$ is acting. The 1st term is the Lorentz force and the 2nd the force due to external potential, and the magnetic term does not appear explicitly. Let us remind ourselves that the minimal electromagnetic interaction is introduced in order to obtain the Lorentz force.

## 6. EPILOGUE

We have collected for convenience a number of formulas associated with the quantized Schrödinger field. In particular, expressions for various physical quantities, such as the number, energy and momentum densities, are given explicitly, rather than going into technical harassment of the Green's function method.

The physical quantities are usually defined as response functions under certain operations, or putting it differently, as generators of certain transformations. Various relations between physical quantities are conveniently given in canonical field theory.

The equal-time commutation relation between a pair of physical quantities $A(\vec{x},t)$ and $B(\vec{x}',t')$, say, calculated in canonical field theory (cf. Appendix A) can be converted to the relation without the equal-time restriction by the help of the identity

$$[A(\vec{x},t),B(\vec{x}',t)]$$

$$= \int_{-\infty}^{\infty} dt'\{T(A(\vec{x},t)\partial_{t'}B(\vec{x}',t'))$$

$$- \partial_{t'}T(A(\vec{x},t)B(\vec{x}',t'))\} \tag{6.1}$$

where T is the chronological ordering, i.e.

$$T(A(\vec{x},t)B(\vec{x}',t')) = \begin{cases} A(\vec{x},t)B(\vec{x}',t') & t>t' \\ B(\vec{x}',t')A(\vec{x},t) & t'>t \end{cases} . \tag{6.2}$$

Such a relation without the equal-time restriction is often called the generalized Ward relation, which essentially shows correct book-keeping in canonical field theory[7]. For example, a number of f-sum rules are derived as a special consequence of Ward relations. As the structure of our physical objects becomes more complicated, it becomes harder to solve fundamental equations exactly and therefore the nature of the approximation used will become unintelligible, unless a correct book-keeping is ensured at every step of approximations.

The form of the generalized Ward relations are particularly suitable for the derivation of relations among various propagators, since they no longer involve the notion of equal-time. The equal-time commutators calculated in Appendix A can thus be put into the generalized Ward form by virtue of eq. (6.1) without difficulty, thereby correct book-keeping among various propagators is always assured.

# APPENDIX A

$$[J_o(\vec{x},t),J_o(\vec{x}',t)] = 0 \tag{A.1}$$

$$[J_o(\vec{x},t),J_j(\vec{x}',t)] = -i\,\frac{\hbar}{m}\,\{\partial_j J_o(\vec{x},t)\delta(\vec{x}-\vec{x}') + J_o(\vec{x},t)\partial_j\delta(\vec{x}-\vec{x}')\} \tag{A.2}$$

$$[J_i(\vec{x},t),J_j(\vec{x}',t)] = -i\,\frac{\hbar}{m}\,\{\partial_j J_i(\vec{x},t)\delta(\vec{x}-\vec{x}') + J_i(\vec{x},t)\partial_j\delta(\vec{x}-\vec{x}')+J_j(\vec{x},t)\partial_i\delta(\vec{x}-\vec{x}')\} \tag{A.3}$$

$$[J_o(\vec{x},t),T_{oo}(\vec{x}',t)] = -i\hbar\{\partial_k J_k(\vec{x},t)\delta(\vec{x}-\vec{x}') + J_k(\vec{x},t)\partial_k\delta(\vec{x}-\vec{x}')\} \tag{A.4}$$

$$\begin{aligned}[J_i(\vec{x},t),T_{oo}(\vec{x}',t)] = -i\hbar\{&\partial_k J_{ki}(\vec{x},t)\delta(\vec{x}-\vec{x}') \\ &+ \frac{1}{2m}\,\psi^\dagger(\vec{x},t)\partial_i(v_o(\vec{x},t)+2V^{ext}(\vec{x}))\psi(\vec{x},t)\delta(\vec{x}-\vec{x}') \\ &+ \frac{1}{2m}\,\partial_i V(|\vec{x}-\vec{x}'|)\psi^\dagger(\vec{x},t)\psi^\dagger(\vec{x}',t)\psi(\vec{x}',t)\psi(\vec{x},t) \\ &+ J_{ki}(\vec{x},t)\partial_k\delta(\vec{x}-\vec{x}') - \left(\frac{\hbar}{2m}\right)^2\partial_k(\partial_k J_o(\vec{x},t)\partial_i\delta(\vec{x}-\vec{x}')) \\ &+ \frac{1}{m}\,T_{oo}(\vec{x},t)\partial_i\delta(\vec{x}-\vec{x}')\}\end{aligned} \tag{A.5}$$

$$[J_o(\vec{x},t),T_{oj}(\vec{x}',t)] = -i\hbar\{\partial_j J_o(\vec{x},t)\delta(\vec{x}-\vec{x}') + J_o(\vec{x},t)\partial_j\delta(\vec{x}-\vec{x}')\} \tag{A.6}$$

$$[J_i(\vec{x},t),T_{oj}(\vec{x}',t)] = -i\hbar\{\partial_j J_i(\vec{x},t)\delta(\vec{x}-\vec{x}')$$

$$+ J_i(\vec{x},t)\partial_j\delta(\vec{x}-\vec{x}')+J_j(\vec{x},t)\partial_i\delta(\vec{x}-\vec{x}')$$

$$+ i\,\frac{\hbar}{2m}\,J_o(\vec{x},t)\partial_i\partial_j\delta(\vec{x}-\vec{x}')\} \qquad \text{(A.8)}$$

where

$$J_{ij}(\vec{x},t) \equiv \left(-i\,\frac{\hbar}{2m}\right)^2 \psi^\dagger(\vec{x},t)(\partial_i-\overleftarrow{\partial}_i)(\partial_j-\overleftarrow{\partial}_j)\psi(\vec{x},t). \qquad \text{(A.9)}$$

## REFERENCES

1. J.M. Lêvy-Leblond, Comm. Math. Phys. 6 (1967) 286 .
2. Z. Koba, G. Takeda, Prog. Theoret. Phys. 3 (1948) 98, 203, 407; Prog. Theoret. Phys. 4 (1949) 60.
3. S.M. Dancoff, Phys. Rev. 55 (1939) 959.
4. Y. Takahashi, "Generating Current and Dynamical Rearrangment of Symmetries". The Ta-You Wu Festschrift: Science of Matter, Ed. by S. Fujita, Gordon and Breach, New York, 1978.
5. Y. Takahashi, Phys. Lett. 67A (1978) 385.
6. W.J. Hurley, Phys. Rev. D3 (1971) 2339, see also ref. 1.
7. See for example, Y. Takahashi, "Canonical field theory and the Ward relations". Positano Symposium on Quantum Field Theory, North-Holland, 1985.

# TUNNELLING THROUGH A BARRIER VIA STOCHASTIC QUANTIZATION

JOSE A. MAGPANTAY
National Institute of Physics
University of the Philippines
Diliman, Quezon City 3004
PHILIPPINES

## A B S T R A C T

We discuss tunnelling through a potential barrier using the language of stochastic quantization. We find that we have to make each collective coordinate (associated with global, continuous symmetries broken by the classical solution) dependent on the extra "time" $\tau$. We derive then the Langevin equations satisfied by the non-zero fluctuation and the collective coordinates and from these derive the distribution function.

## I. INTRODUCTION

Before I met Professor Gyo Takeda, I only know of him from the Proceedings of the 1978 Tokyo conference on High Energy Physics. I heard from so many physicists that the Tokyo conference was one of the best and I thought that this is a good reflection on him being the organizer. Then I met Prof. Takeda for the first time when he visited the University of the Philippines on December '83 and I got to know him better when I visited Tohoku University six months later (both visits were sponsored by the Japan Society for the Promotion of Science). I realized during my visit why the Tokyo Conference was successful: (i) the Japanese government gives full support for the basic sciences, (ii) Professor Takeda really takes care of his guests.

I also learned in my visit that Professor Takeda gives a serious concern for the condition of physics and physicists in a country like the Philippines. This concern is not even shown by some science policy makers of my country. Thus, when Prof. Yoshiyuki Kawazoe invited me to write a paper to honor Professor Takeda on his sixtieth birthday, I readily agreed. I just hope that this contribution measures up to the man being honored, Prof. Gyo Takeda.

## II. TUNNELLING VIA EUCLIDEAN PATH-INTEGRALS:

Let us consider a system with potential function $V[\phi]$ shown by (simplified configuration space diagram) Figure 1.

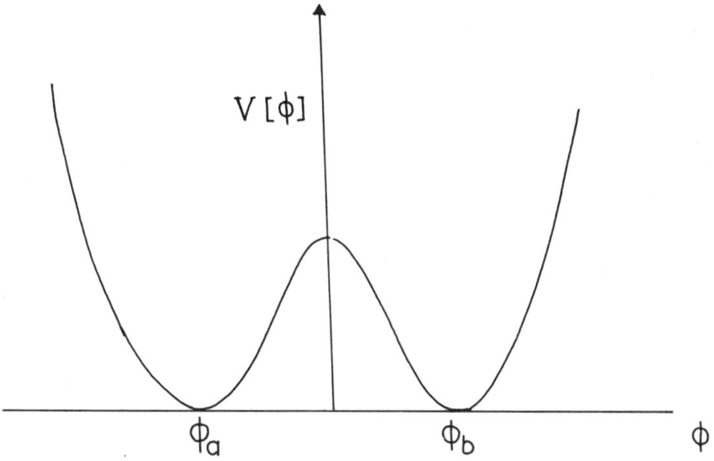


Figure 1: Simplified Potential Function in Configuration Space.

There is a potential barrier between the points labeled by $\phi_a$ and $\phi_b$. Tunnelling between these configurations will occur if there is a Euclidean field configuration $\tilde{\phi}(\vec{x},x_4)$ such that $\tilde{\phi}(\vec{x},x_4 = 0) = \phi_a$ and $\tilde{\phi}(\vec{x},x_4 = T) = \phi_b$, i.e., the Euclidean configuration interpolates between the two configurations. The tunnelling amplitude is given by

$$<\phi_a|e^{-HT}|\phi_b> = \int \prod_i d\omega_i [df]\{\det[<\frac{\delta\tilde{\phi}}{\delta\omega_i}|\frac{\delta\tilde{\phi}}{\delta\omega_j}>]\}^{\frac{1}{2}}$$

$$\times \exp. -S_E[\tilde{\phi}] - \tfrac{1}{2}\int f \left.\frac{\delta^2 S_E}{\delta\phi^2}\right|_{\phi=\tilde{\phi}} f dx \qquad (II.1a)$$

$$= \int \prod_i d\omega_i \{\det[<\frac{\delta\tilde{\phi}}{\delta\omega_i}|\frac{\delta\tilde{\phi}}{\delta\omega_j}>]\}^{\frac{1}{2}} \exp.\{-S_E[\tilde{\phi}]\}$$

$$\times [\det' \left.\frac{\delta^2 S_E}{\delta\phi^2}\right|_{\phi=\tilde{\phi}}]^{\frac{1}{2}} \qquad (II.1b)$$

For clarity, we would like to explain the determinants that appear in (II.1a, II.1b). The determinant involving the zero modes $\frac{\delta\tilde{\phi}}{\delta\omega_i}$ is only an ordinary matrix determinant and $<\frac{\delta\tilde{\phi}}{\delta\omega_i}\ \frac{\delta\tilde{\phi}}{\delta\omega_j}>$ representing the inner product between the two zero modes, i.e.,

$$<\frac{\delta\tilde{\phi}}{\delta\omega_i}\bigg|\frac{\delta\tilde{\phi}}{\delta\omega_j}> = \int dx\ \frac{\delta\tilde{\phi}}{\delta\omega_i}\ \frac{\delta\tilde{\phi}}{\delta\omega_j}\ . \tag{II.2}$$

The determinant that appears with $\frac{\delta^2 S}{\delta\phi^2}\Big|_{\phi=\tilde{\phi}(x,\omega)}$ is a functional determinant and the prime denotes the removal of the zero modes. Thus, it is a mathematically well-defined object modulo ultra-violet divergencies.

The problem we will be discussing is how to describe this tunnelling through a barrier in the language of stochastic quantization. The rational for asking such a question is that the Fokker-Planck formulation prove the equivalence of the stochastic method to Euclidean quantum theory in a general manner. Furthermore, explicit verification of the equivalence of the two have been shown perturbatively. Thus, it is only natural to ask if tunnelling can also be discussed in the language of stochastic quantization.

## III. TUNNELLING VIA STOCHASTIC QUANTIZATIONS

Before we discuss tunnelling, we first give a brief review of stochastic quantization. In the stochastic method, one introduces an extra time $\tau$ and a white noise $\eta(x,\tau)$ and starts with the Langevin equation

$$\frac{\partial\phi}{\partial\tau} = - \left.\frac{\delta S_E}{\delta\phi}\right|_{\phi(x,\tau)} + \eta(x,\tau). \qquad \text{(III.1)}$$

The solution to (III.1) is represented by $\phi_\eta(x,\tau)$ and correlation function are calculated via the following prescription

$$<0|\phi(x_1)\ldots.\phi(x_n|0> = \lim_{\tau\to\infty} <\phi_\eta(x_1,\tau)\ldots.\phi_\eta(x_n,\tau)>. \qquad \text{(III.2)}$$

The average on the right hand side of (III.2) is over the white noise , which means we use the following relations

$$<\eta(x,\tau)\eta(x',\tau')>_\eta = 2\delta(x-x')\delta(\tau-\tau'), \qquad \text{(III.3a)}$$

$$<\eta(x_1,\tau_1)\ldots.\eta(x_n,\tau_n)>_\eta = 0, \text{ for n odd,} \qquad \text{(III.3b)}$$

$$<\eta(x_1,\tau_1)\ldots.\eta(x_n,\tau_n)>_\eta = 2\{\delta(x_1-x_2)\delta(\tau_1-\tau_2) \ldots.\delta(x_{n-1}-x_n)\delta(\tau_{n-1}-\tau_n)+ \text{permutations}\} \qquad \text{(III.3c)}$$

In practice, one only verifies (III.2) perturbatively. A non-perturbative proof of (III.2) relies on the equivalent Fokker-Planck formulation of the Langevin equation (III.1). The relationship of the two is defined by

$$<\phi_\eta(x_1,\tau)\ldots.\phi_\eta(x_n,\tau)>_\eta \equiv \int[d\phi(x)]\phi(x_1) \ldots\phi(x_n)P[\phi,\tau]. \qquad \text{(III.4)}$$

Equation (III.4) can be used to derive the Fokker-Planck Hamiltonian that governs the evolution of $P[\phi,\tau]$ given the Langevin equation (III.1). The answer is

$$H_{FP} = \int dx \frac{\delta}{\delta\phi(x)}\left[\frac{\delta}{\delta\phi(x)} + \frac{\delta S_E}{\delta\phi(x)}\right] . \tag{III.5}$$

If $H_{FP}$ is positive, semi-definite ($E>0$) with a unique ground state, it follows that

$$\lim_{\tau\to\infty} P[\phi,\tau] \sim e^{-S_E[\phi]} \tag{III.6}$$

Equation (III.4) and (III.6) then prove the result (III.2).

Now, suppose the system is somewhere near the vicinity of $\phi_a$ (see Figure 1). Its random motion due to the white noise cannot take it through the potential barrier to arrive at $\phi_b$, unless of course $\eta$ is large enough to dominate the restoring force $-\frac{\delta S_E}{\delta\phi}$. We will not consider this possibility for it will be difficult to put a handle on a noise term that is large. Besides, a large noise term will invalidate perturbation theory which has its own range of validity (so far, explicit verification of the equivalence of the stochastic method to Euclidean quantum theory in only perturbative).

Since it is the Euclidean solution $\tilde{\phi}(x)$ which interpolates between $\phi_a$ and $\phi_b$, we want then to find a solution of (III.1) such that

$$\lim_{\tau\to\infty} \langle\phi_\eta(x,\tau)\rangle_\eta = \tilde{\phi}(x). \qquad \text{(III.7)}$$

Furthermore, we must be able to derive the tunnelling amplitude (II.2a,b) from the equivalent Fokker-Planck formulation of the Langevin equation.

The perturbative solution of (III.1) which we label as $\phi_\eta^{p.t.}(x,\tau)$ cannot satisfy (III.7). This can verified by considering point mechanics with the potential

$$V(x) = \tfrac{1}{4}(x^2-1)^2. \qquad \text{(III.8)}$$

The Langevin equation is

$$\frac{\partial x}{\partial \tau} = -\frac{\partial^2 x}{\partial t^2} - x + x^3 + \eta. \qquad \text{(III.9)}$$

The perturbative solution (III.9) involves the greens function of the operator $\frac{\partial}{\partial\tau} + \frac{\partial^2}{\partial t^2} + 1$ . Since the noise term always appear in odd powers, then $\langle x_\eta^{p.t.}(t,\tau)\rangle_\eta = 0$. However, we know that the tunnelling solution is $\bar{x}(t) = + \tan h[\frac{1}{\sqrt{2}}(t - t_0)]$, thus verifying that the perturbative solution cannot satisfy (III.7).

The simplest way to satisfy (III.7) is to employ the background field decomposition in (III.1), then linearize the Langevin equation. This means, we write

$$\phi(x,\tau) = \tilde{\phi}(x) + f(x,\tau) \qquad \text{(III.10)}$$

where $\tilde{\phi}$ is the Euclidean solution. The fluctuation $f(x,\tau)$ must then satisfy

$$\frac{\partial f}{\partial \tau} = - \left.\frac{\delta^2 S_E}{\delta \phi^2}\right|_{\phi=\tilde{\phi}} f + \eta. \tag{III.11}$$

If there are no continuous, global symmetries broken by $\tilde{\phi}$, the operator $\frac{\partial}{\partial \tau} + \left.\frac{\delta^2 S_E}{\delta \phi^2}\right|_{\phi=\tilde{\phi}}$ has a well-defined Greens function g and the solution of (III.11) is given by

$$f(x,\tau) = \int dx_1 d\tau_1 g(x-x_1 \tau-\tau_1)\eta(x_1,\tau_1). \tag{III.12}$$

Equations (III.10) and (III.12) plus the white noise character of $\eta$ give $\lim\limits_{\tau\to\infty} \langle\phi_\eta(x,\tau)\rangle_\eta = \tilde{\phi}$, which is the tunnelling solution. Thus, the ansatz given by the decomposition (III.10) and the Langevin equation (III.11) is apparently the correct approach in treating tunnelling through a barrier. This will be further supported by the fact that we can derive the tunnelling amplitude by considering the Fokker-Planck equation corresponding to the Langevin equation (III.11).

The distribution function is given by the Fokker-Planck equation

$$- \frac{\partial P}{\partial \tau} = \int dx \frac{\delta}{\delta f(x)}\left[\frac{\delta}{\delta f(x)} + \left.\frac{\delta^2 S_E}{\delta \phi^2}\right|_{\phi=\tilde{\phi}} f\right]P. \tag{III.13}$$

The solution to this equation is given by the Wiener integral

$$P[f,\tau;\ f_o,\tau{=}0] = \int\limits_{\text{end points}} [d\eta(x,\tau)]\ \exp.\{-\tfrac{1}{4}\int dx d\tau\ \eta^2(x,\tau)\}. \tag{III.14}$$

Converting this to an integration with respect to the fluctuation f, we find

$$P[f,\tau;\ f_o,\tau{=}0] = \int\limits_{\text{end points}} [df(x,\tau)]\ \det[\frac{\partial}{\partial\tau} + \frac{\delta^2 S_E}{\delta\phi^2}\Big|_{\phi=\tilde{\phi}}] \tag{III.15}$$
$$\times\ \exp.\{-\tfrac{1}{4}\int dx d\ (\frac{\partial f}{\partial\tau} + \frac{\delta^2 S_E}{\delta\phi^2}\Big|_{\phi=\tilde{\phi}}\ f)^2.$$

Using the stationary phase approximation (which gives the exact result because of quadratic Fokker-Planck action), we find

$$P = \{\det[e^{a\tau} - e^{-a\tau}]\delta(x-x')\}^{\frac{1}{2}}$$
$$\times\ \exp.\{-\ \tfrac{1}{2}\int dx[f-e^{-a\tau}f_o][1+e^{-a\tau}G^{-1}(a\tau)]a \tag{III.16}$$
$$\times\ [f-e^{-a\tau}f_o]\}.$$

where $a = \frac{\delta^2 S_E}{\delta\phi^2}\Big|_{\phi=\tilde{\phi}}$ and $G(a\tau) = e^{a\tau} - e^{-a\tau}$. In the limit $\tau\to\infty$, the distribution function relaxes to the equilibrium value

$$P[f(x)] \sim \exp.\{-\tfrac{1}{2}\int dx\ f\ \frac{\delta^2 S_E}{\delta\phi^2}\Big|_{\phi=\tilde{\phi}}\ f\}. \tag{III.17}$$

This is just the usual path integral result.

Equation (III.15) clearly shows that the Fokker-Planck distribution is well-defined in the case when the classical configuration $\tilde{\phi}$ does not break any continous, global symmetries. In this case, the operator $\frac{\partial}{\partial\tau} + \frac{\delta^2 S_E}{\delta\phi^2}\Big|_{\phi=\tilde{\phi}}$

does not have square-integrable zero modes. In transforming the path-integral from the white noise $\eta$ to the fluctuation $f(x,\tau)$ (see equation (III.15)), the Jacobian is a determinant of a well-defined operator.

However, in general, classical configurations do break some global symmetries of the theory: In this case, the operator $\frac{\partial}{\partial\tau} + \frac{\delta^2 S_E}{\delta\phi^2}\Big|_{\phi=\tilde{\phi}}$ will have zero modes given by $\frac{\delta\omega(x,\omega)}{\delta\omega_i}$ [to make the zero modes square-integrable, we first put the system inside the "box" defined by $[0<\tau<L]$. The Langevin equation (III.11) must then be supplemented by the condition.

$$\int_0^L d\tau \int dx \frac{\delta\tilde{\phi}(x,\omega)}{\delta\omega}\eta(x,\tau) = 0, \tag{III.18}$$

i.e., the zero modes orthogonal to the source $\eta$. This condition will contradict the white noise character of $\eta$. To see this, consider

$$\int dx_1 d\tau_1 \frac{\delta\tilde{\phi}(x_1,\omega)}{\delta\omega}\eta(x_1,\tau_1) \int dx d\tau \frac{\delta\tilde{\phi}}{\delta\omega}(x,\omega)\,\eta(x,\tau)$$

By equation (III.18), this must be zero. However, if we take the average over the white noise $\eta$, we find

$$\int_0^L d\tau \int dx \frac{\delta\tilde{\phi}(x,\omega)}{\delta\omega}\frac{\delta\tilde{\phi}(x,\omega)}{\delta\omega}.$$

This is not necessarily zero. Thus, we see that the existence of zero modes contradicts the white noise character of $\eta$ if the starting point is the Langevin equation (III.11). Also, the Fokker-Planck equation

(III.13) does not follow from the Langevin equation (III.11) because the derivation of (III.13) is based on the white noise character of $\eta$.

In terms of the "path-integral" approach, the problem arises in transforming the "path-integral" over the white noise to the fluctuation f. The Jacobian is zero because of the existence of zero modes for the operator $\frac{\partial}{\partial\tau} + \frac{\delta^2 S_E}{\delta\phi^2}\Big|_{\phi=\tilde{\phi}}$.

How do we properly take into account the zero modes? Actually, the Euclidean path-integral method gives us a hint. Equation (II-2a) suggests that the distribution function depends not only on the non-zero fluctuations but also on the collective coordinates. Furthermore if (II.2a) is derivable from the stochastic method, the Fokker-Planck distribution must relax into the Euclidean result, i.e.,

$$\lim_{\tau\to\infty} P[f(x),\omega;\tau] \sim \exp.\{-S_E[\tilde{\phi}] - \tfrac{1}{2}\int dx f \frac{\delta^2 S_E}{\sigma\phi^2}\Big|_{\phi=\tilde{\phi}} f + \tfrac{1}{2}\, \underset{i,j}{\mathrm{tr}}\, \ln < \frac{\delta\tilde{\phi}}{\delta\omega_i}\Big|\frac{\delta\tilde{\phi}}{\delta\omega_j} >\} \tag{III.19}$$

Since the distribution function depends on the collective coordinate $\omega$, there must be a corresponding Langevin equation for it. To derive the appropriate Langevin equations, let us substitute the background field decomposition

$$\phi(x) = \tilde{\phi}(x,\omega(\tau)) + f(x,\tau), \tag{III.20}$$

in (III.1). Note that in (III.20), we are assuming that the collective coordinate is "time" dependent. Furthermore, $f(x,\tau)$ is purely non-zero fluctuation.

Corresponding to the decomposition (III.20), let us also separate out the noise term into

$$\eta(x,\tau) = \eta_f(x,\tau) + c_i(x)\eta_i(\tau), \qquad \text{(III.21)}$$

when $\eta_f(x,\tau)$ is the noise term for the non-zero fluctuation $f(x,\tau)$ and $\eta_i(\tau)$ is the corresponding noise for the collective coordinate $\omega_i(\tau)$. The coefficients $c_i(x)$ are determined by imposing the condition

$$\int [d\eta(x,\tau)] e^{-\frac{1}{4}\int dx d\tau\, \eta^2(x,\tau)} = \left(\int [d\eta_f(x,\tau)] e^{-\frac{1}{4}\int dx d\tau\, \eta_f^2(x,\tau)}\right) \times \left(\int [d\eta_i(\tau)] e^{-\frac{1}{4}\int d\tau\, \eta_i^2(\tau)}\right). \qquad \text{(III.22)}$$

This condition insures that both noises $\eta_f$ and the $\eta_i$'s are white and no mixing between the two. To arrive at the right-hand-side of (III.22), the $c_i(x)$ must satisfy

$$\int dx\, c_i(x) c_j(\lambda) = \delta_{ij}, \qquad \text{(III.23a)}$$

$$\int dx d\tau\, c_i(x)\eta_i(\tau)\eta_f(x,\tau) = 0, \text{ for each } i. \qquad \text{(III.23b)}$$

A set $c_i(x)$ that will satisfy above is

$$c_i(x) = \left\langle \frac{\delta\phi}{\delta\omega_i} \middle| \frac{\delta\phi}{\delta\omega_j} \right\rangle^{\frac{1}{2}} \frac{\delta\phi}{\delta\omega_j}, \qquad \text{(III.24)}$$

where $< \frac{\delta\tilde{\phi}}{\delta\omega_i} | \frac{\delta\tilde{\phi}}{\delta\omega_j} >^{-\frac{1}{2}}$ represents the matrix square-root of the inverse of $< \frac{\delta\tilde{\phi}}{\delta\omega_i} | \frac{\delta\tilde{\phi}}{\delta\omega_j} >$.

Substituting (III.20) and (III.21) in (III.1), we find the following Langevin equations

$$\frac{\partial f}{\partial \tau} = - \frac{\delta^2 S_E}{\delta\phi^2}\Big|_{\phi=\tilde{\phi}} f + \eta_f, \qquad \text{(III.25a)}$$

$$\frac{\partial \omega_i}{\partial \tau} = < \frac{\delta\tilde{\phi}}{\delta\omega_i} | \frac{\delta\tilde{\phi}}{\delta\omega_j} >^{-\frac{1}{2}} \eta_j(\tau). \qquad \text{(III.25b)}$$

These equations tell us that the collective coordinates satisfy a Langevin equation with a multiplicative white noise but without a restoring force. Thus, the collective coordinates are doing a pure random walk motion which is consistent with the fact that they are related to zero modes which lie at the trough of the action in configuration space. The non-zero fluctuation f on the other hand has a linear restoring force. The consistency condition in solving (III.25a) does not pose a problem now as it is trivially satisfied when we imposed (III.23b) to make the noises $\eta_f(x,\tau)$ and $\eta_i(\tau)$ white.

The Fokker-Planck Hamiltonian corresponding to the Langevin equations is[4]

$$H_{FP} = \int dx \frac{\delta}{\delta f(x)}\Big[\frac{\delta}{\delta f(x)} + \frac{\delta^2 S_E}{\delta\phi^2}\Big|_{\phi=\tilde{\phi}} f\Big] \qquad \text{(III.26)}$$

$$+ \frac{\delta}{\delta\omega_i} \{G_{ij}G_{jk}[\frac{\delta}{\delta\omega_k} + \frac{\delta tr\ln G}{\delta\omega_k}]\}$$

where

$$G_{ij} = < \frac{\delta\tilde{\phi}}{\delta\omega_i} \Big| \frac{\delta\tilde{\phi}}{\delta\omega_j} >^{-\frac{1}{2}} \tag{III.27}$$

The zero mode of this Hamiltonian is

$$\begin{aligned} \psi_0 &\sim \exp.\{-\tfrac{1}{2} \int dx\, f \frac{\delta^2 S_E}{\delta\phi^2}\Big|_{\phi=\tilde{\phi}} f - \mathrm{tr}\, \ell n\, G\} \\ &= \det\, [<\frac{\delta\tilde{\phi}}{\delta\omega} \frac{\delta\tilde{\phi}}{\delta\omega}>]^{\frac{1}{2}} \exp. -\tfrac{1}{2} \int dx\, f \frac{\delta^2 S_E}{\delta\phi^2}\Big|_{\phi=\tilde{\phi}} f\}. \end{aligned} \tag{III.28}$$

This is proportional to the equilibrium distribution function usually arrived at via the path-integral method (see (III.29)).

The Langevin equation (III.25), for the collective coordinates does not have a restoring force. Yet, we find from (III.28) that the collective coordinate eventually attains equilibrium with distribution exp. $\{-\mathrm{tr}\, \ell n\, G\}$. How is this possible? The answer lies in the fact that (III.25b) is a Langevin equation with multiplicative noise, thus it can always be expressed as a Langevin equation with an additive noise with a restoring force.[5] This makes the collective coordinates also attain equilibrium with the distribution function given in (III.28).

Thus, we have shown that tunnelling through a barrier can also be derived in the language of stochastic quantization.

REFERENCES:


1. For an introduction and guide to the original literature, see the following reviews: C. Bernard, Talk Presented at the Canadian Mathematical Society Summer Institute (1979) and R.J. Crewther, in "Facts and Prospects of Gauge Theories," proceedings of the XVII Schladming Conference on Nuclear Physics edited by P. Urban (Springer, Berlin, 1978) [Acta Phys. Austriaca Suppl. 19 XXX (1978)].

2. G. Parisi and Y.S. Wu, Scientia Sinica 24 48 (1981).

3. R. Courant and P. Hilbert, "Methods of Math. Physics (Interscience, New York, 1953), Vol. 1, p. 355.

4. J.A. Magpantay and D.M. Yanga, "Continuum Non-Perturbative Effects Via Stochastic Quantization" National Institute of Physics Preprint 85/11.

5. See for example, K.L.C. Hunt and J. Ross, Journal of Chemical Physics 75 976 (1981.)

# MICROCANONICAL QUANTIZATION AND ITS APPLICATIONS*

A. Iwazaki

Nuclear Science Division

Lawrence Berkeley Laboratory

University of California

Berkeley, California 94720

ABSTRACT

We discuss microcanonical quantization and its interesting properties and explain our recent proposal for a numerical application.

Microcanonical quantization[1,2] is derived from the analogy between statistical mechanics and functional quantization of fields. As is well-known, the standard functional quantization in Euclidean space may be recognized as classical canonical ensemble averages. With the Euclidean action $S_0(\phi)$ of a field $\phi$, the functional quantization leads to

$$\langle f(\phi)\rangle = \frac{1}{Z_0}\int D\phi\, f(\phi)e^{-S_0(\phi)} \quad , \quad Z_0 \equiv \int D\phi\, e^{-S_0(\phi)} \qquad (1)$$

---

*This work was supported by the Director, Office of Energy Research, Division of Nuclear Physics of the Office of High Energy and Nuclear Physics of the U.S. Department of Energy under Contract DE-AC03-76SF00098.

where $f(\phi)$ is a physical quantity. This equation may be rewritten by inserting auxiliary variables, $P_i$

$$<f(\phi)> = \frac{1}{Z}\int DPD\phi\ f(\phi)e^{-\beta H(P,\phi)} \quad , \quad Z \equiv \int DPD\phi^{-\beta H(P,\phi)}$$

$$H \equiv \sum_{i=1}^{N} \frac{P_i^2}{2} + \frac{S_0(\phi)}{\beta} \tag{2}$$

where $\beta$ is an arbitrary constant and the index i runs over the dynamical degrees of freedom of the field $\phi$ (e.g. lattice sites if we adopt the lattice regularization). The equation (2) gives the canonical ensemble average in a system with Hamiltonian $H(P,\phi)$ and a temperature $\beta^{-1}$; $P_i$ and $\phi_i$ are canonical conjugates to each other. We thus understand that functional quantization is equivalent to taking the canonical ensemble average. Then, we are led naturally to introduce a microcanonical ensemble $\delta(E - H)$ and to take an average as

$$<f(\phi)>_m = \frac{1}{Z_m}\int DPD\phi\ f(\phi)\delta(E - H) \quad , \quad Z_m \equiv \int DPD\phi\ \delta(E - H) \quad . \tag{3}$$

This is just the one as we call microcanonical quantization.[1] This method has been used[3] numerically in lattice gauge theories and the numerical results agree well with Monte-Carlo results of (1).

To formulate the method more rigorously, we need to take a regularization (hereafter, we use the lattice regularization) and to take a finite volume (=V). The equivalence between $<f>$ and $<f>_m$ has been proved[1] perturbatively in the limit of the infinite volume,

$$\lim_{V\to\infty} \langle f\rangle = \lim_{V\to\infty} \langle f\rangle_m \tag{4}$$

In this proof of the perturbative equivalence, the energy in eq. (3) should be taken as $E = N/\beta$, where N is the number of dynamical degrees of freedom ($DPD\phi \equiv \sum_{i=1}^{N} dP_i d\phi_i$). It goes without saying that perturbation theory is reliable in the limit of small coupling constant, g. Hence, the relation of $E = N/\beta$ holds only around $g = 0$. To show the equivalence in eq. (4) for arbitrary g, it can be argued that we need the following relation,

$$E = \frac{N}{2\beta} + \frac{1}{Z_0}\int D\phi\; e^{-S_0} S_0/\beta \tag{5}$$

where $N/2\beta$ comes from the kinetic energy of $\sum_{i=1}^{N} P_i^2/2$. The origin of this relation is easy to understand by taking the canonical average of $H(P,\phi)$. In addition, we need some further assumptions in the proof of the equivalence for general g, which are physically reasonable and rather weak, see ref. 1. The eq. (5) may be replaced with

$$E = \frac{N}{2\beta} + \frac{1}{\beta}\langle S_0\rangle_m \tag{6}$$

This is a consistency condition, which is useful in numerical calculations.

In short, the microcanonical quantization is performed by use of the microcanonical ensemble and with the consistency condition in eq. (6). We can check explicitly its validity by examining several solvable problems in the statistical mechanics. For example, we have computed the magnetizations in a classical Heisenberg spin model by use of a mean field approximation. That calculation is a bit tedious and we need some tricks, even though that problem can be solved simply in the canonical ensemble average.

Now, we comment on some interesting features of the microcanonical quantization. One of them is the absolute convergence of a perturbation series,[4] which is defined by expanding $\delta(E - H)$ with respect to the coupling constant g by regarding E as an independent constant: We are considering a situation where the action $S_0$ has a form as $S_0 = S_f + gS_I$ (scalar field theory) or $S_0 = S/g^2$ (lattice gauge theory). Hence, eq. (6) is used to determine E in terms of g after we sum the perturbation series of $\langle S \rangle_m$. It is amazing that, by using this perturbation theory, we can obtain, in principle, not only asymptotic properties as $g \to 0$, but also general properties for arbitrary g of the quantum field theory. We note that the standard perturbation series in the functional quantization does not converge even if the theory has ultraviolet and infrared cutoffs; this implies that the standard perturbation theory reveals only asymptotic properties of the theory as $g \to 0$.

Another interesting feature lies in its numerical applications[3,5] where one postulates the "ergodicity" of the energy surface $E = H$ in phase space $\{P_i, \phi_i\}$. We shall consider a U(1) lattice gauge theory as an example. The action of the lattice gauge theory is a function of the link variables, $U_i = e^{i\theta_i}$ and the inversed coupling constant, $\beta$: $S_0 = \beta S(U)$. If we identify $\beta$ in eq. (2) with this inversed coupling constant, the Hamiltonian becomes $H = \sum_{i=1}^{N} P_i^2/2 + S(U)$ and the microcanonical ensemble average is given by

$$\langle f\rangle_m = \frac{1}{Z_m}\int \prod_{i=1}^{N} dP_i d\theta_i \delta(E - H) f(\theta) \quad , \quad 2\pi \geq \theta \geq 0 \quad . \tag{7}$$

We note that $N/2\beta$ can be computed as

$$\frac{N}{2\beta} = \left\langle \sum_{i=1}^{N} P_i^2/2 \right\rangle_m \tag{8}$$

This means that each kinetic energy contributes $1/2\beta$ to the energy (equipartition law); $\beta^{-1}$ plays the role of temperature.

Now, let us assume ergodicity of the classical system described by the Hamiltonian $H(P,\theta)$. This assumption allows us to compute $\langle f\rangle_m$ by solving the Hamliton's equations,

$$\frac{d\theta_i}{d\tau} = \frac{\partial H}{\partial P_i} \quad , \quad \text{and} \quad \frac{dP_i}{d\tau} = -\frac{\partial H}{\partial \theta_i} \tag{9.a}$$

and by taking an average over time $\tau$,

$$\lim_{T\to\infty} \frac{1}{T}\int_0^T f(\theta(\tau))dz = \langle f\rangle_m \quad . \tag{9.b}$$

The ergodicity is expected to hold in almost all nonlinear systems for sufficiently large number of the dynamical degrees of freedom. Therefore, in the microcanonical quantization, problems in the quantum theory are reduced to problems in the corresponding classical theory. However, if the system has some explicit symmetries, the ergodicity doesn't hold. In our case of the U(1) lattice gauge theory, we have local gauge symmetries, which allow us to reduce some irrelevant cyclic variables. Hence, in order to apply the above procedure, we must reformulate the microcanonical quantization in terms of independent dynamical variables. This leads us to solve the Hamilton's equation for all variables $(\theta, P)$ under initial conditions such as $P_i = 0$ for all i. Then, the irrelevant variables $P_i'$ vanish at any time so that we must replace N in the left hand side of eq. (8) with the number of independent degrees of freedom, $N_{in}$. As an example, we have performed the numerical calculation of $\langle S\rangle_m$ vs $\beta$ using this method. We adopt a lattice size of $4^4$ and S as

$$S = \sum_p (1 - \frac{1}{2}(U_p + U_p^+)) \quad , \quad U_p \equiv \sum_{i \in p} e^{i\theta_i} \tag{10}$$

where P indicates a plaquette.

First, we start with random configuration of $\theta_i$ and with $P_i = 0$ for all i, and thermalize the system. The temperature $\beta^{-1}$ and the internal energy $\langle S\rangle_m$ are measured by taking their averages over large enough time interval T,

$$\frac{N_{in}}{2\beta} = \frac{1}{T}\int_0^T \sum_{i=1}^N P_i^2/2 \, d\tau \quad \text{and} \quad \langle S\rangle_m = \frac{1}{T}\int_0^T S(\theta(\tau))d\tau \quad , \tag{11}$$

where $\tau = 0$ is not the initial time, but a time after the system is thermalized. Next, we decrease (or increase) the temperature. To do so, we use our recently proposed method[5] for changing the temperature. That is, we put a friction term into the Hamilton's equations

$$\frac{d\theta_i}{d\tau} = \frac{\partial H}{\partial P_i} \quad \text{and} \quad \frac{dP_i}{d\tau} = -\frac{\partial H}{\partial \theta_i} + c'P_i \tag{12}$$

with $c' \equiv C_0/\sum_{i=1}^{N} P_i^2$, where $C_0$ is a constant (hereafter, we only discuss the case of decreasing the temperature, that is, $C_0 > 0$). Due to this friction term, the energy of the system decreases as $\frac{dH}{d\tau} = -C_0$, and the temperature also decreases continuously in time. Hence, if we measure $\beta$ and $\langle S\rangle_m$ as a function of the time, $\tau$, we can determine the relation between $\beta$ and $\langle S\rangle_m$. We depict the result in Fig. 1, which agree well with Monte-Carlo results. We conclude that the microcanonical simulation described above works well.

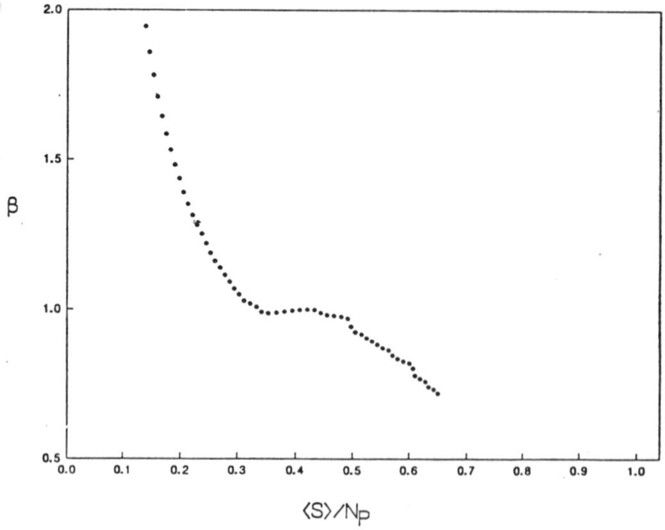


Fig. 1.

$N_p$ = # of plaquette
$C_0 = 2$
$T = 2500$

As a final comment on an interesting feature of the microcanonical simulation, we shall briefly state an application of our proposal to identify the order of phase transitions. The distinction of first order phase transitions with second order ones has been performed mainly in Monte-Carlo simulations by examining, for example, the continuity of the internal energy. However, the discontinuity of a critical point of the first order is smoothed out by finite volume effects. On the other hand, it is possible in the microcanonical simulation to identify the first order phase transitions even in a finite volume.

According to the microcanonical simulation, we can find the internal energy of supercooled (or superheated) metastable states. Namely, for a system with sufficiently small volume, its internal energy $\langle S\rangle_m$ becomes a multi-valued function of the temperature $\beta^{-1}$ beyond a critical point of a first order. One branch of the function corresponds to the energy of the supercooled (or superheated) state. This is because, even beyond the critical point, such metastable states cannot decay into a stable state; surface energy effects between the two phases prevent the decay of the metastable states in a system with sufficiently small volume. The multi-valuedness is revealed by examining the so-called S shaped curve[6] in the $\langle S\rangle_m$ vs. $\beta$ plane. On the other hand, the multi-valuedness is not expected around a critical point of second order. The reason is that the internal energy is smooth at this critical point.

Taking an action,

$$S = \sum_p \left\{(1 - \frac{1}{2}(U_p + U_p^+)) + (1 - \frac{1}{2}(U_p^2 + U_p^{+2}))\right\} \tag{13}$$

we depict these typical S shaped curves in Fig. 2 which was obtained by using our method[5] in eq. (12) for decreasing the temperature; curve B is from ref. 6 where a different method for decreasing the temperature was used in the microcanonical simulation. Curves A and C correspond to different choices of the parameter $C_0$. The reason for a difference between these two curves around the critical point is that the extent to which the system is supercooled depends on the rate (=$C_0$) at which the temperature is decreased. We find that our curves, A and C, show more clearly S shape behavior of the system than does the curve B obtained by another method. Therefore, it is possible to obtain fairly clear S shaped curves by choosing an appropriate friction coefficient, $C_0$, and hence to identify unambiguously the first order phase transition.

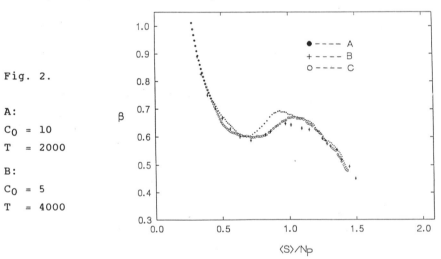

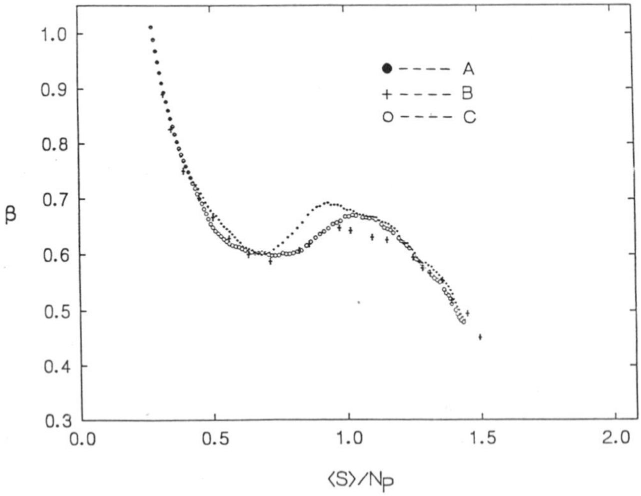


Fig. 2.

A:
$C_0$ = 10
T = 2000

B:
$C_0$ = 5
T = 4000

ACKNOWLEDGEMENT

This paper was written at the occasion of Prof. G. Takeda's 60th anniversary. The author greatly appreciates his constant encouragement and educational attitudes. He also expresses his gratitude to Dr. F. Klinkhamer for reading through the manuscript. The last part of this paper was done in collaboration with Dr. Y. Morikawa.

This work was supported by the Director, Office of Energy Research, Division of Nuclear Physics of the Office of High Energy and Nuclear Physics of the U.S. Department of Energy under Contract DE-AC03-76SF00098.

## REFERENCES

1. A. Iwazaki, Phys. Lett. 141B (1984) 342, LBL preprint LBL-19909 to be published in Phys. Rev. D.
2. D.J.E. Callaway and A. Rahman, Phys.Rev. Lett. 49 (1982).
3. M. Creutz, Phys. Rev. Lett. 50 (1983) 1411.
   J. Polony and H.W. Wyld, Phys. Rev. Lett. 51 (1983) 2257.
   J. Polony, H.W. Wyld, J.B. Kogut, J. Shigemitsu and D.K. Sinclair, Phys. Rev. Lett. 53 (1984) 644.
   J.B. Kogut, J. Polony, H.W. Wyld and D.K. Sinclair, Phys. Rev. Lett. 54 (1985) 1980.
4. A. Iwazaki, Phys. Lett. 159B (1985) 348.
5. Y. Morikawa and A. Iwazaki, LBL preprint LBL-20175 to be published in Phys. Lett. B.

# QUASISTABLE SOLITONS WITH AXIAL BARYON NUMBER

Z. F. Ezawa
Department of Physics, Tohoku University, Sendai 980, Japan
and
T. Yanagida
Institute of Physics, College of General Education, Tohoku University, Sendai 980, Japan

*We study the nonlinear σ model coupled with ρ mesons. This model possesses two kinds of topological charges. One topological charge is absolutely conserved and identified with the baryon number, while the other is not conserved because of the ρ meson instanton effect and identifiable with the axial baryon number. We find a quasi-stable soliton carrying the axial baryon number whose mass is less than 2 GeV. It is suggested that a possible candidate for this quasi-stable soliton is the ι (1.4) meson.*

The nonlinear σ model gives rise to stable static classical solutions called Skyrmions [1], when a term which is quartic in the time derivative is added. Skyrmions carry a topological quantum number B to be identified with the baryon number [2]. In particular, the lowest energy state will be nothing but nucleons [3].

In this paper we analyze a nonlinear σ-model coupled with massive vector mesons. There is a very successful scheme for this purpose proposed Weinberg [4]. As was recently emphasized [5], his Lagrangian has a local SU (2) gauge symmetry implicitly. Therefore, the relevant coset space is not simply $\{SU(2)_L \times SU(2)_R\}/SU(2)$ but rather

$\{SU(2)_L \times SU(2)_R \times SU(2)_H\}/SU(2)$, where $SU(2)_L \times SU(2)_R$ stand for the chiral groups while $SU(2)_H$ denotes the local symmetry just mentioned. Consequently, we have two kinds of topological charges B and $\tilde{B}$, since

$$\Pi_3 \{SU(2)_L \times SU(2)_R \times SU(2)_H/SU(2)\} = Z(B) + Z(\tilde{B}). \tag{1}$$

The first charge B should be identified with the baryon number while the second charge $\tilde{B}$ may be associated with the axial baryon number. However, just as in the minimal nonlinear $\sigma$ model, there is no stable soliton in the Weinberg chiral Lagrangian. It is necessary to introduce higher derivative terms (Skyrme terms) in order to stabilarize topological solitons. In this modified model, baryons are shown to be absolutely stable. We also find quasi-stable solitons carrying the topological charge $\tilde{B}$ in a wide range of parameters of the model. Here, the instability is due to $\rho$-meson instanton effects. We estimate the mass and the life-time of the lightest quasi-stable soliton. The mass of the quasi-soliton, if exist, is expected to be less than 2 GeV, and the life-time to be around ten times longer than typical hadronic ones. It would be very intriguing to identify such an object with the $\iota$ meson [6].

We start with the Weinberg chiral Lagrangian [4]:

$$\begin{aligned} L_0 = & \frac{f_\pi^2}{4} \mathrm{Tr}(\partial_\mu U \partial^\mu U^\dagger) \\ & + a f_\pi^2 \mathrm{Tr}\{\hat{V}_\pi - \frac{1}{2i}(\partial_\mu \xi \cdot \xi^\dagger + \partial_\mu \xi^\dagger \cdot \xi)\}^2 \\ & - \frac{1}{2g^2} \mathrm{Tr}\, \hat{F}_{\mu\nu} \hat{F}^{\mu\nu} \end{aligned} \tag{2}$$

with

$$\hat{F}_{\mu\nu} = \partial_\mu \hat{V}_\nu - \partial_\nu \hat{V}_\mu - i[\hat{V}_\mu, \hat{V}_\nu], \tag{3}$$

$$U = \xi^2 = \exp(i2\pi/f_\pi). \tag{4}$$

Here, $\pi = \frac{1}{2}\sum_{a=1}^{3} \pi^a(x)\tau^a$ and $\hat{V}_\mu = \frac{1}{2}\sum_{a=1}^{3} \hat{V}_\mu^a \tau^a$ are the pion and the $\rho$ meson fields, respectively; $f_\pi = 93$ MeV is the pion decay constant. This Lagrangian is phenomenologically very successful. The coupling constant and the mass of the $\rho$ mesons are given by

$$g_{\rho\pi\pi} = \frac{a}{2} g, \quad m_\rho^2 = a g^2 f_\pi^2. \tag{5}$$

The choice a=2 leads to the universality and the KSRF relation [8].

However, there is no stable soliton in the Weinberg chiral Lagrangian (2) just as in the minimal nonlinear $\sigma$ model. It is necessary to assume Skyrme terms which prevent the solitons from shrinking to zero-size. We thus add the following two terms to $L_0$:

$$L_1 = \frac{1}{32e^2}\mathrm{Tr}[\partial_\mu U\cdot U^\dagger,\ \partial_\nu U\cdot U^\dagger]^2$$
$$+\frac{1}{32c^2}\mathrm{Tr}[\hat{V}_\mu - \frac{1}{2i}(\partial_\mu\xi\cdot\xi^\dagger + \partial_\mu\xi^\dagger\cdot\xi),\ \hat{V}_\nu - \frac{1}{2i}(\partial_\nu\xi\cdot\xi^\dagger + \partial_\nu\xi^\dagger\cdot\xi)]^2. \qquad (6)$$

They are the unique terms with four derivatives that lead to a positive Hamiltonian second order in time derivatives.

Let us set $U = \xi^2 = 1$ in Lagrangian (2) with (6), since we are only concerned about solitons carrying the charge $\tilde{B}$. It follows that

$$L = -\frac{1}{2g^2}\mathrm{Tr}\ \hat{F}_{\mu\nu}^2 + \frac{m^2}{g^2}\mathrm{Tr}\ \hat{V}_\mu^2 + \frac{1}{32c^2}\mathrm{Tr}[\hat{V}_\mu,\ \hat{V}_\nu]^2, \qquad (7)$$

with $m^2 = ag^2f_\pi^2$. We discuss properties of quasi-stable solitons of this massive Yang-Mills model. Here, the last term in (7) is needed for the stability of solitons. This Lagrangian is regarded as the Yang-Mills theory where the Higgs-Kibble mechanism is operating. Indeed, substituting

$$\hat{V}_\mu \equiv S^\dagger V_\mu S + iS^\dagger\partial_\mu S, \quad SS^\dagger = 1 \qquad (8)$$

we obtain

$$L = -\frac{1}{2g^2}\mathrm{Tr}\ F_{\mu\nu}^2 + \frac{m^2}{g^2}\mathrm{Tr}\ (D_\mu S)(D^\mu S)^\dagger + \frac{1}{32c^2}\mathrm{Tr}[D_\mu S\cdot S^\dagger,\ D_\nu S\cdot S^\dagger]^2, \qquad (9)$$

where the field $S(x)$ is considered as the Higgs field which is "frozen" with $SS^\dagger = 1$.

The topological property is understood in the both forms of the Lagrangians (7) and (9) as follows. Let us consider the weak coupling limit $g^2 \to 0$ with $m/g$ being fixed in (7). The finiteness of the Hamiltonian requires $\hat{F}_{\mu\nu} = 0$, or

$$\hat{V}_\mu = iS^\dagger\partial_\mu S. \qquad (10)$$

In this limit the Lagrangian may be approximated by the substitution of $\hat{V}_\mu$ with (10) in (7).

$$L' = \frac{m^2}{g^2}\mathrm{Tr}(\partial_\mu S\partial^\mu S^\dagger) + \frac{1}{32c^2}\mathrm{Tr}[\partial_\mu S\cdot S^\dagger,\ \partial_\nu S\cdot S^\dagger]^2. \qquad (11)$$

Equivalently, switching off the gauge field $V_\mu$ in (9) also yields (11). This Lagrangian has precisely the same form as the Skyrme model. Soliton solutions are easily found by making the Skyrme anzatz:

$$S = \exp\{iF(r)\hat{x}^a \tau^a\}. \tag{12}$$

The soliton with the boundary conditions

$$F(r) = \pi \quad \text{at } r = 0, \quad F(r) \to 0 \quad \text{as } r \to \infty \tag{13}$$

carries one unit of the topological charge ($\tilde{B} = 1$).

Here, the absolutely conserved topological current is given by

$$\tilde{B}^0_\mu = \frac{1}{24\pi^2} \varepsilon_{\mu\nu\rho\sigma} \mathrm{Tr}[\partial^\nu S \cdot S^\dagger \partial^\rho S \cdot S^\dagger \partial^\sigma S \cdot S^\dagger].$$

but this is not a gauge invariant quantity when $g^2 \neq 0$. The gauge invariant current is defined by [9, 7]

$$\tilde{B}_\mu = \frac{1}{24\pi^2} \varepsilon_{\mu\nu\rho\sigma} \mathrm{Tr}[D^\nu S \cdot S^\dagger D^\rho S \cdot S^\dagger D^\sigma S \cdot S^\dagger + \frac{3}{2} iF^{\nu\rho} D^\sigma S \cdot S^\dagger], \tag{14a}$$

where

$$D_\mu S = \partial_\mu S - iV_\mu S.$$

This current $\tilde{B}_\mu$ can also be expressed in terms of the physical $\rho$ meson field $\hat{V}_\mu$ as

$$\tilde{B}_\mu = \frac{i}{24\pi^2} \varepsilon_{\mu\nu\rho\sigma} \mathrm{Tr}[\hat{V}^\nu \hat{V}^\rho \hat{V}^\sigma - \frac{3}{2} i\hat{F}^{\nu\rho} \hat{V}^\sigma], \tag{14b}$$

where a use was made of (8). Unlike $\tilde{B}^0_\mu$, the gauge invariant current is not conserved:

$$\partial^\mu \tilde{B}_\mu = \frac{1}{16\pi^2} \mathrm{Tr}\, F_{\mu\nu} \tilde{F}^{\mu\nu}. \tag{15}$$

Hence, the associated charge $\tilde{B} = \int d^3x\, \tilde{B}_0$ is time dependent; solitons carrying charge $\tilde{B}$ are unstable.

To see a nature of the topological charge $\tilde{B}$, we introduce the coupling between quarks and the $\rho$ mesons:

$$L_q = \Sigma \bar{q} \gamma^\mu (i\partial_\mu + V_\mu) q. \tag{16}$$

Then, the axial baryon number current $\tilde{J}_\mu = \bar{q}\gamma_\mu \gamma_5 q$ is anomalous:

$$\partial_\mu \tilde{J}^\mu = \frac{1}{16\pi^2} \mathrm{Tr}\, \tilde{F}_{\mu\nu} F^{\mu\nu}. \tag{17}$$

From (15) and (17) the topological charge $\tilde{B}$ is found to be related to the axial baryon number.

We have argued that topological excitations in Lagrangian (7) are unstable for any finite values of $g^2$. We now estimate the mass and the life-time of such a quasi-stable soliton.

In principle we may calculate the mass with an arbitral accuracy by finding the local minimum of the Hamiltonian in the total functional space of $\hat{V}_\mu$, which we assume to be realized by $\hat{V}^0_\mu$. Then, the decay probability is obtained by minimizing the Eucledian action in the functional subspace of $\hat{V}_\mu$ which consists of the field configurations connecting $\hat{V}_\mu = \hat{V}^0_\mu$ and $\hat{V}_\mu = 0$. However, this is not practical. In this paper we assume a one-parameter family of field configurations which connect the Skyrme solution (10) with the vacuum $\hat{V}_\mu = 0$. By minimizing the Hamiltonian and the action with respect to this parameter we estimate the mass and the decay probability. Thus, the true mass of the quasi-soliton is smaller than our estimation. To check the accuracy of our calculation, we evaluate the vacuum tunnelling amplitude in the pure Yang-Mills theory by our one-parameter family of field configurations and compare our result with that of instanton effects.

Let us consider a one-parameter family of field configurations of the following form:

$$\hat{V}_\mu = i\theta S^\dagger \partial_\mu S. \tag{18}$$

When $\theta = 1$, it represents the Skyrme solution (10) with (12), while when $\theta = 0$ it just represents the vacuum ($\hat{V}_\mu = 0$). From the Lagrangian (7) together with this ansatz, we find the mass of the soliton to be

$$M(\theta) = \frac{F_0^2}{16} \int d^3x\, \mathrm{Tr}\,(\partial_i S \partial_i S^\dagger) - \frac{1}{32e_0^2} \int d^3x\, \mathrm{Tr}\,[\partial_i S \cdot S^\dagger,\ \partial_j S \cdot S^\dagger]^2 \tag{19}$$

with

$$F_0 = \frac{4m}{g}\theta, \tag{20 a}$$

$$\frac{1}{e_0^2} = \frac{16}{g^2}\theta^2 \{(1-\theta)^2 + \frac{g^2}{16c^2}\theta^2\}, \tag{20 b}$$

Here, we use the spherical anzatz (12) and change the integration variable from r to the dimensionless one by $R = e_0 F_0 r$:

$$M(\theta) = \frac{F_0}{e_0} \{I(F) + J(F)\} \tag{21}$$

with

$$I(F) = 4\pi \int_0^\infty R^2 dR \frac{1}{8} \{(\partial_R F)^2 + 2\frac{\sin^2 F}{R^2}\}, \tag{22 a}$$

$$J(F) = 4\pi \int_0^\infty R^2\, dR \frac{1}{2} \frac{\sin^2 F}{R^2} \{\frac{\sin^2 F}{R^2} + 2(\partial_R F)^2\}. \tag{22 b}$$

For an arbitral value of the parameter $\theta$, the mass $M(\theta)$ is first minimized in the functional space of $F(R)$ with the boundary condition (13). This problem has already been solved and we know that [3]

$$I(F) + J(F) = 36.5 \tag{23}$$

We next minimize $M(\theta)$ with respect to $\theta$.

It is easy to find the following properties of $M(\theta)$. If

$$g^2 > \frac{2}{3} c^2, \tag{24 a}$$

there is only one minimum point at $\theta = 0$, which is the vacuum solution. In this case there is no soliton. If

$$g^2 < \frac{2}{3} c^2, \tag{24 b}$$

there are two local minimum points at $\theta = \theta_+$ and $\theta = 0$, and one local maximum point at $\theta = \theta_-$; the point $\theta = \theta_+$ corresponds to a quasi-soliton. The mass values at these points $\theta_\pm$ are given by

$$M(\theta_+) \approx 146 \frac{m}{g^2} \frac{g}{c}, \tag{25 a}$$

$$M(\theta_-) \approx 86.5 \frac{m}{g^2}, \tag{25 b}$$

for $g^2 \ll \frac{2}{3} c^2$, and

$$\theta_\pm = \frac{1}{6(1+\gamma)} [5 \pm \sqrt{1-24\gamma}], \quad \gamma = \frac{g^2}{16c^2} \tag{26}$$

We have depicted the function $M(\theta)$ in Fig. 1a. It is interesting to see that the maximum value $M(\theta_-)$ is insensitive to the parameter $g/c$. The mass of a quasi-stable soliton, if exist, should be smaller than $M(\theta_-)$.

We next evaluate the decay probability of this quasi-stable soliton. The soliton is assumed to decay by following the path of field configurations given by (18). Instead of evaluating the Eucledian action, we may estimate the probability for the "coordinate" $\theta$ to tunnel from $\theta = \theta_+$ to $\theta = 0$ in Fig. 1a. For this purpose we introduce the time-dependence to the parameter $\theta$. Substituting (18) into the Lagrangian density (7) and integrating over the x-space, we find the Lagrangian for the coordinate $\theta(t)$:

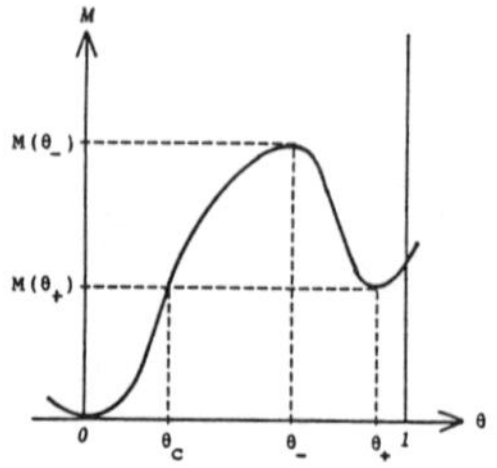


Fig. 1a

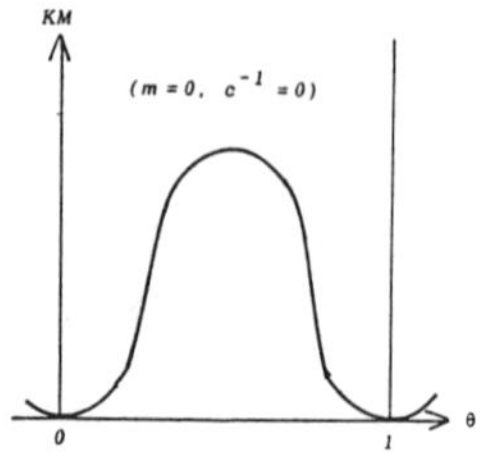


Fig. 1b

*Fig. 1a) The mass function $M(\theta)$ for field configurations (18) which connects a soliton solution with the vacuum. The local minimum at $\theta = \theta_+$ corresponds to the quasi-stable soliton.*

*Fig. 1b) The potential barrier between two topologically distinct vacuums in a pure Yang-Mills theory with field configurations (18).*

$$L = \frac{K}{2}\, \dot{\theta} - M(\theta), \tag{27}$$

where $M(\theta)$ is given by (21) and

$$K = \frac{g^2}{2} \int d^3 x \; \mathrm{Tr}\,(\partial_i S \partial^i S^\dagger)$$

$$= \frac{32}{g^2} \frac{1}{e_0 F_0} I(F)\,. \tag{28}$$

Here, $e_0 F_0$ and $I(F)$ have been defined in (20) and (22). It is a simple exercise to calculate the tunnelling probability in the Lagrangian (27). In the WKB approximation the decay probability P is obtained as

$$P = \exp\left(-2\int_{\theta_c}^{\theta_+} d\theta\, [2K(\theta)\, \{M(\theta) - M(\theta_+)\}]^{1/2}\right) \tag{29}$$

where $\theta_c$ is the point at which $M(\theta_c) = M(\theta_+)$; see, Fig. 1a.

We now wish to estimate the accuracy of our approximation. For this purpose we calculate the vacuum tunnelling probability in the pure Yang-Mills theory by our method. This is given simply by setting $m = 0$ and $c^{-1} = 0$ in the previous formulas; see, Fig. 1b. The vacuum tunnelling probability $P_v$ is found as

$$P_v = \exp\{-2\int_0^1 d\theta\, \sqrt{2KM}\}$$

$$= \exp\{-\frac{32}{3g^2} \sqrt{IJ}\}$$

$$= \exp\{-1.23 \frac{16\pi^2}{g^2}\}. \tag{30}$$

The tunnelling probability due to an instanton is known to be $\exp(-16\pi^2/g^2)$. It will be allowed to guess that our estimation to the action is at most 1.23 times bigger than the true value.

Finally, let us apply our results to the $\rho$-meson physics, where $m^2 = 2g^2 f_\pi^2 = (770 \text{ MeV})^2$. The local maximum of $M(\theta)$ is determined to be

$$M(\theta_-) \approx 1.9 \text{ GeV}.$$

Therefore, the mass of the quasi-soliton, if exist, should be less than 2 GeV. It would be very intriguing to identify this quasi-soliton with the $\tau$ meson. If it is the case, the mass of the $\tau$ meson [6] implies $g^2/c^2 \approx 0.18$. With this value of c, the decay probability is calculable from (29) to be order of $10^{-1}$. Therefore, we expect that the decay width of this quasi-soliton is around ten times smaller than typical widths of hadrons, which is consistent with the observed width of $\tau$ meson. However, in order to justify our conjecture, it is necessary to examine various quantum numbers of this soliton which are still lacking.

However, if $g^2/c^2 > 2/3$ in our model, there may be no chance to have such a quasi-stable soliton. In this sence, the Weinberg-Salam model is more promising in producing quasi-solitons since $g_W \ll g_\rho$. In this case the field $V_\mu$ and S stand for the weak guage and Higgs bosons, respectively. Using $m \approx 80$ GeV and $g_W^2/4\pi \approx 1/25$, we find from the formula (25) that $M \approx 23(g_W/c_W)$ TeV.

We are very grateful to Professor G. Takeda for many stimulating conversations on the Skyrme model.

## References

1. T. H. R. Skyrme, Proc. Roy. Soc. A260 (1961) 127.
2. A. P. Balachandran, V. P. Nair, S. G. Rajeev and A. Stern, Phys. Rev. Letters 49 (1982) 1124; Phys. Rev. D27 (1983) 1153.
   E. Witten, Nucl. Phys. B223 (1983) 422; B223 (1983) 433.
3. G. S. Adkins, C. R. Nappi and E. Wittern, Nucl. Phys. B228 (1983) 552.
4. S. Weinberg, Phys. Rev. 166 (1968) 1568.
5. M. Bando, T. Kugo, S. Uehara, K. Yamawaki and T. Yanagida, Phys. Rev. Letters 54 (1985)

1215.

6. Particle Data Group, Phys. Letters 111B (1982) 1.
7. J. M. Gipson, Nucl. Phys. B231 (1984) 365;
   E. D'Hoker and E. Farhi, Nucl. Phys. B241 (1984) 109.
8. J. J. Sakurai, Gurrent and Mesons, (Univ. Chicago Press, Chicago, 1969).
9. J. Goldstone and F. Wilczek, Phys. Rev. Letters 47 (1981) 986.

# Violation of Decoupling Theorem and Uniqueness of Induced Topological Term in Three Dimensions

O. Abe* and K. Ishikawa

Department of Physics, Hokkaido University Sapporo, Japan

*Production Engineering Research Laboratory, Hitachi Ltd., Yokahama, Japan

Abstract

Appelquist-Carrazone's decoupling theorem is shown not to work in three dimensional fermion theory in which the fermion couples with the external electromagnetic field. Instead of complete decoupling of heavy fermion, topological term is emerged in the effective Lagrangian. The magnitude of the topological term is uniquely determined regardless of the fermion's interaction with a scalar field. The exact low energy theorem can be written. The other cases such as the charged boson and the two dimensional fermion theory are also examined. The violation of decoupling theorem occurs only in the three dimensional fermion theory.

A response of a charged particle against an external electric field should be getting smaller as the particle is getting heavier. The force is proportional to the electric charge. Thus the acceleration is inversely proportional to the mass from Newton's law. The particle does not move if it has the infinite mass.

The above-mentioned inverse proportionalily in the classical mechanics is extended easily to that of the quantum mechanics. The classical equation of motion hold in quantum mechanics.

In quantum field theory, the equation of motion is that of the wave field. The relation between the mass and the response may not be so simple. However the physical effect of the large mass field has been shown to be negligible in the low energy region. The theorem is very general. It was proved by Appelquist and Carrazone[1] and called as decoupling theorem.

The decoupling theorem has been proved in 4-dimensions. An investigation in other dimensions have been lacking. We shall study the problem in two and three dimensions in this paper. Low energy phenomena in these dimensions can be studied experimentally in condensed matter physics. Hence there could be some direct physical implications.

We study the response of the system of charged heavy particle against the external electromagnetic field. In the large mass limit of the charged fermion fields, the effective Lagrangian does not vanish in three dimensions contrary to the naive expectation based on the classical picture, but a topological Chern-Simon term remains. The magnitude of the topological term is determined by the number of charged fermions alone. The charged boson does not contribute in the limit. In two dimensions, the decoupling of the massive field is complete.

I. 3 dimensions.

We shall describe the theory in three dimensions first and the one in two dimensions later. The Lagrangian density[2] which we study is given by

$$\mathcal{L} = \bar{\psi}\gamma^{\mu}(i\hbar\partial_{\mu}+eA_{\mu})\psi - m\bar{\psi}\psi + g\bar{\psi}\psi\phi - \frac{1}{2}(\partial_{\mu}\phi)^2 - \frac{1}{2}\mu^2\phi^2. \qquad (1)$$

The charged fermion $\psi$ couples to the external electromagnetic field $A_{\mu}$ and the neutral scalar field $\phi$ in three dimensions. The gamma matrices $\gamma_{\mu}$'s are given by 2×2 and a standard representations is given by the Pauli matrices.

We study the effective Lagrangian in terms of $A_{\mu}$ after the fermion and the neutral scalar field freedoms are integrated. Diagramatically, this corresponds to investigate Feynmann diagrams with the external $A_{\mu}$ lines. Let $I_A$, $n_F$, $n_B$, and $n_v$ be the number of external $A_{\mu}$ lines, the number of internal fermion lines, the number of internal boson lines, and the number of vertices. The degree of divergences of the diagram which has $I_A$ external $A_{\mu}$ lines and $n_B$ internal boson lines in D-dimensions are given by

$$D(n_B+n_F-n_v+1) - n_F - 2n_B = D - I_A + (D-4)n_B \quad , \qquad (2)$$

One feature in low dimensions (D=2 or 3) is represented by the fact that the increasing number of internal boson lines makes the diagram less divergent, since the coupling g has a positive dimension of mass (super-renormalizable).

The diagram is expected to be convergent if $D-I_A+(D-4)n_B$ is negative. Then it behaves as the negative power of the fermion's mass if the external momenta are much smaller than the fermion's mass. For more rigorous argument, we shall use Weinberg's theorem.

The interesting case in 3 dimensions is limited to the following cases : $I_A=2$, $n_B=0,1$ and $I_A=3$, $n_B=0$.

(1) $I_A = 2$, $n_B = 0,1$ (Fig. 1.a, b)

Transversality makes us possible to write the amplitude as

$$\pi_{\mu\nu}(p) = \varepsilon_{\mu\nu\rho} p^\rho \pi_1(p^2) + (p^2 g_{\mu\nu} - p_\mu p_\nu)\pi_2(p^2) \tag{3}$$

using two invariant amplitudes $\pi_1(p^2)$ and $\pi_2(p^2)$. The degrees of divergences of $\pi_1(p^2)$ and $\pi_2(p^2)$ are reduced by one and two respectively from those given in Eq.(2). Thus $\pi_1(p^2)$ from $I_A=2$ $n_B=0$ diagram has a vanishing degree of divergence. The following amplitude $\pi_1(p^2)$ remains constant in the large fermion mass limit.

$$\begin{aligned}\pi_{\mu\nu}(p) &= -\int \frac{d\ell}{(2\pi)^3} \mathrm{Tr}\,[\gamma_\mu \frac{\not\ell+m}{\ell^2+m^2} \gamma_\mu \frac{\not\ell-\not p+m}{(\ell-p)^2-m^2}] \\ &= \varepsilon_{\mu\nu\rho} p^\rho \pi_1(p^2) + (p^2 g_{\mu\nu} - p_\mu p_\nu)\pi_2(p^2),\end{aligned}$$

where

$$\pi_1(p^2) = \frac{-i}{4\pi}\{1 + \frac{1}{12}\frac{p^2}{m^2} + \cdots\} \overset{m\to\infty}{=} \frac{-1}{4\pi}$$

$$\pi_2(p^2) = \frac{1}{12\pi m}\{1 + \frac{1}{10}\frac{p^2}{m^2} + \cdots\} \overset{m\to\infty}{=} 0. \tag{4}$$

The $\pi_2(p^2)$ has always a negative degree of divergence and vanishes in the large fermion mass limit. It may be worth while to emphasize that the boson exchange diagrams such as those of Fig.1(b) vanish in this limit.

For a rigorous proof of the above arguments, Weinberg's theorem[3] is used. We show it in the particular example given in Fig.1(b). The amplitude $\pi_{\mu\nu}(p)$ is written as

$$-\int \frac{dq}{(2\pi)^3} \frac{d\ell}{(2\pi)^3} \mathrm{Tr}\,[\gamma_\mu \frac{1}{\not{q}-m} \frac{1}{\not{q}+\not{\ell}-m} \gamma_\nu \frac{1}{\not{q}+\not{\ell}-\not{p}-m} \frac{1}{\ell^2-\mu^2} \frac{1}{\not{q}-\not{p}-m}] \tag{5}$$

Thus $\varepsilon_{\mu\nu\alpha}\pi_1(p)$ term comes from

$$-\int \frac{dq}{(2\pi)^3} \frac{d\ell}{(2\pi)^3} \{\mathrm{Tr}[\gamma_\mu \frac{\not{q}+m}{q^2-m^2} \frac{\not{q}+\not{\ell}+m}{(q+\ell)^2-m^2} \gamma_\nu (-)\gamma_\alpha \frac{1}{(q+\ell-p)^2-m^2} \frac{1}{\ell^2-\mu^2} \frac{\not{q}+m}{(q-p)^2-m^2}] + \mathrm{Tr}[\gamma_\mu \frac{\not{q}+m}{q^2-m^2} \frac{\not{q}+\not{\ell}+m}{(q+\ell)^2-m^2} \gamma_\nu \frac{\not{q}+\not{\ell}+m}{(q+\ell-p)^2-m^2} \frac{1}{\ell^2-\mu^2} (-)\gamma_\alpha \frac{1}{(q-p)^2-m^2}]\} \tag{6}$$

Since the above equation includes $\varepsilon_{\mu\nu\alpha}\pi_1(p^2)$ term, we can find the behavior of $\pi_1(p^2)$ from them. By rescaling the integration variables $q_\mu$ and $\ell_\mu$ with the dimensionless vectors $t_\mu$ and $s_\mu$ as

$$q_\mu = mt_\mu \ , \qquad \ell_\mu = ms_\mu \ , \tag{7}$$

we see that the first term of Eq.(9) becomes to $\frac{1}{m}$ I in the large m limit, where

$$I = \int \frac{dt}{(2\pi)^3} \frac{ds}{(2\pi)^3} \{\mathrm{Tr}[\gamma_\mu \frac{\not{t}+1}{t^2-1} \frac{\not{t}+\not{s}+1}{(t+s)^2-1} \gamma_\nu\gamma_\alpha \frac{1}{(t+s)^2-1} \frac{1}{s^2} \frac{\not{t}+1}{t^2-1}]\} \tag{8}$$

Now we apply Weinberg's theorem in order to prove the convergence of I. It is easy to see from the power counting that any subdiagrams described in Fig.(2) are convergent. Thus the integral I is convergent.
The second term of the Eq.(6) and other similar diagrams such as those given in Fig.(3) behave also as inverse power of the large mass.
The proof is the same as the above-mentioned one and is not presented here. Since the vertex correction is finite, the application to higher order diagrams is straight forward.

The amplitude $\pi_1(p^2)$ is exactly known in the large mass limit. Conversely, in the low momentum limit the amplitude is given by the same formula.

(2) $I_A = 3,\ n_B = 0$ (Fig. 1.c)

The diagram which has a vanishing degree of the divergence vanishes from Bose symmetry in U(1) theory which we are discussing. But it does not vanish in non-Abelian theory in whcih $A_\mu$ and $\psi$ have an internal degree of freedom. Combined amplitude of this term with the previous one is invariant under the non-Abelian gauge transformation.

II. 2 dimensions

The degree of divergence of the above theory in two dimensions is given by $2-I_A-2n_B$. The degree of divergence of $\pi_2(p)$ vanishes for $I_A=2$ and $n_B=0$ and is negative for other cases. The computation of the amplitude is straightforward and the result is

$$\frac{1}{6\pi}(g_{\mu\nu}p^2-p_\mu p_\nu)\frac{p^2}{m^2}\{1+0(\frac{p^2}{m^2})\} \tag{9}$$

for p<<m.

This vanishes as m goes to infinity. Hence the decoupling of the heavy field holds in two dimensions.

III. Charged scalar

The degree of divergence given in Eq.(2) are now replaced with

$$D - I_A + (\frac{D}{2} - 3)n_v \tag{10}$$

for a theory with $\phi^3$ coupling, and

$$D - I_A + (D - 4)n_v \tag{11}$$

for a theory with $\phi^4$ coupling. In the above equations $n_v$ is the number of vertices of the scalar self-interaction. In three dimensions, the degree of divergence takes its maximum value, 1, when $I_A$ and $n_v$ are equal to 2 and 0 respectively.

Since the anti-symmetric tensor $\varepsilon_{\mu\nu\rho}$ does not emerge in the boson theory, contrary to the fermion theory, there is only one invariant amplitude in two point function. The degree of the divergence of $\pi_2(p)$ is less than that of $\pi_{\mu\nu}(p)$ by two. Thus the degree of divergences of $\pi_2(p)$ is always negative. The amplitude vanishes if the mass of the changed scalar is infinite.

The more rigorous proof is given by using Weinberg's theorem. However since we have described it already in the fermionic theory, and the proof is essentially same, we neglect to describe it here.

Two dimensional bosonic theory is essentially the same as the three dimensional bosonic theory. The heavy field decouples in the low energy phenomena.

We have shown that the three dimensional fermion theory is very special. The heavy fermion field bears a nonvanishing term even in the large mass limit, contrary to the decoupling theorem. The term in the effective action has an universal property in a sence that it is irrelevant to the details of the fermion interactions and is given by

$$\frac{e^2}{4\pi} \int \varepsilon_{\mu\nu\rho} A^{\mu} \partial^{\nu} A^{\rho} d^3x \quad . \tag{12}$$

This is Chern-Simon term[4] which possess a topological nature.

The statistics played an important role in generating the induced topological term. Anti-symmetric tensor $\varepsilon_{\mu\nu\rho}$ can appear in the theory of Fermi field, but does not appear in the theory of only Bose field. In one of the previous papers[5], one of the present authors has claimed the connection between the quantized Hall effect and the induced topological term of the Dirac theory. If the electrons would have been a boson, the quantized Hall effect may not have occured.

If the fermion is not elementary but has a spatial structure, it has an anomolous magnetic moment or other higher moments. The induced current in this case should be different from the one of the local field, and has additional terms from the higher moments. These terms which come from the higher moments are proportional to the higher derivatives of the vector potential. The term which is proportional to the magnetic moment is proportional to the first derivative of the electric field, for example. Thus in the low energy limit with respect to the external $A_{\mu}$ lines, the induced electric current agrees with the one obtained before. For the constant electric field, the formula is exact even if the charged particle has a spatial structure.

Acknowledgements

It is a great pleasure to dedicate this article to Professor G. Takeda on the occasion of his sixtieth birthday.

References

1. T. Appelquist and J. Carrazone, Phys. Rev. D11, 2856 (1975).
2. The theory in 2+1 dimensions without scalar field has been investigated from a different context by : A Redlich, Phys. Rev. Lett. 52, 18 (1984); A. Niemi and G. Semenoff, Phys. Rev. Lett. 51, 2077 (1983); K. Ishikawa, Phys. Rev. Lett. 53, 1615 (1984).
3. S. Weinberg, Phys. Rev., 118, 838 (1960).
4. R. Jackiw, "Topological Investigations of Quantized Gauge Theories," in Proceedings of Les Houches 1983 Summer School; S. Deser, R. Jackiw, and S. Templeton, Phys. Rev. Lett., 48, 975 (1982); Ann. Phys. (NY) 140, 379 (1982); J. Schonfeld, Nucl. Phys. B185, 157 (1981).
5. K. Ishikawa, Phys. Rev. Lett. 53, 1615 (1984), Phys. Rev. D31, 1432 (1985); R. Jackiw, Phys. Rev. D29, 2375 (1984).

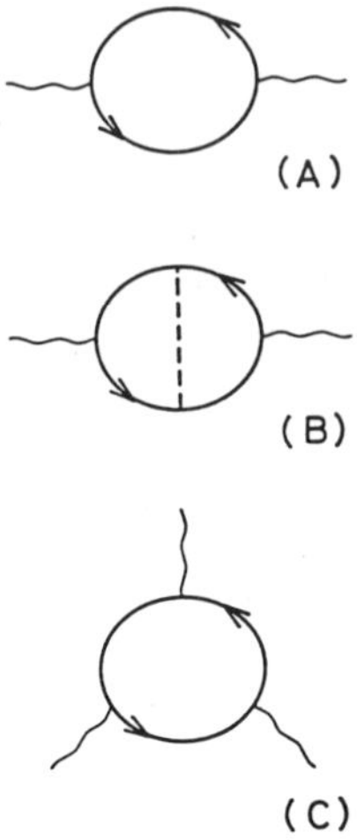


Fig. 1

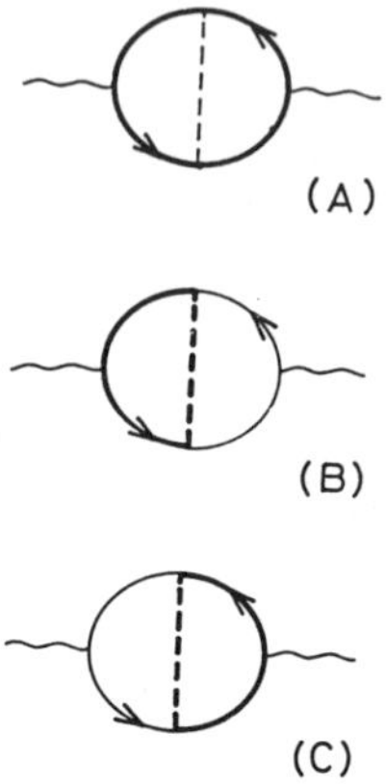


Fig. 2

Fig. 1. The diagrams which have vanishing or positive degrees of divergence in three dimensions. The wavy lines stand for the external electromagnetic field, the dotted line stands for the scalar field and the dashed line denotes the fermi field.

Fig. 2. Three subgraphs (darkened lines) of the graph Fig.1(b) are shown. The degrees of the divergence of three diagrams are negative.

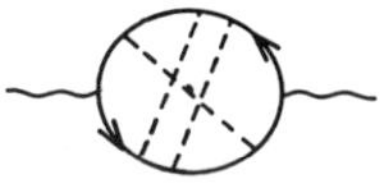

Fig. 3

Fig. 3. An example of the higher order diagrams, which has a negative degree of divergence.

# THERMALLY INDUCED BROKEN SYMMETRIES

H. Umezawa, H. Matsumoto and N. Yamamoto
Department of Physics
The University of Alberta
Edmonton, Alberta T6G 2J1, Canada

ABSTRACT

Thermally induced broken symmetries such as the supersymmetry and the Lorentz symmetry do not necessarily require zero mass Goldstone particles. Some intuition and mathematics are presented behind this controversial subject.

## 1. INTRODUCTION

It is our great pleasure to dedicate this paper to Dr. Gyo Takeda in celebration of his sixtieth birthday. Nearly four decades ago, one of the authors (H. Umezawa) started his research career with a calculation of fourth order Compton scattering in the presence of several kinds of charged particles. Dr. Takeda, being a collaborator of Dr. Z. Koba, was an expert in this very complex type of calculation and H.U. visited Dr. Takeda several times in order to seek his advice. H.U. was then fascinated by his precise, original and stimulating arguments based on beautiful intuitions. They became good friends and H.U. was able to learn very much from Dr. Takeda. Recalling the intuitive and mathematical nature of his arguments, in this paper we try to present the intuition and mathematics behind the controversial subject of thermally induced broken symmetries.

Since the appearance of an ordered state is a result of the spontaneous breakdown of a symmetry and since the ordered states tend to appear at low temperature, it is usually supposed that lowering temperature may cause spontaneous breakdown of symmetries. From this viewpoint, it is rather surprising that breakdown of some symmetries are thermally induced. There are two well-known examples of thermally induced broken symmetries[1,2]; one is the Lorentz symmetry while another is the supersymmetries.

Using the terminology of thermo field dynamics, I. Ojima[1] suggested that these broken symmetries are caused by the presence of quanta of negative energy (i.e. quanta of the tilde field) which are holes in the thermally excited particles. A detailed analysis of thermally induced broken symmetry is now in the process of development. A remarkable and puzzling aspect of thermally induced broken symmetries are that the spontaneous breakdown of symmetry occurs even without interaction.

The first question may be to ask if these symmetries under consideration are really broken and if so, is the breaking spontaneous? This question has already caused a certain amount of controversy, which depends on the formalism being employed. In the case of the Lorentz symmetry, it may be said that its breakdown is an intrinsic one caused by the presence of a special coordinate system associated with a thermal reservoir. Since the spontaneous breakdown of symmetries is a concept defined within the framework of quantum field theory and since thermo field dynamics (TFD)[3] very faithfully adopts the structure of the quantum field theory, it is interesting to analyse the thermally induced broken symmetries in terms of thermo field dynamics for equilibrium systems. According to the terminology of quantum field theory, a spontaneous breakdown of symmetry means that the vacuum state does not have the symmetries as the Heisenberg equations (or as the Langrangian). Applying this definition to TFD, we find that the Lorentz symmetry is spontaneously broken because the Heisenberg equations are Lorentz invariant while the thermal vacuum is not. A similar situation appeared also in the case of supersymmetry (SUSY). Using the usual trace procedure for the thermal average, one finds a non-vanishing order parameter for the supersymmetric models at finite temperature. This led A. Das and M. Kaku[2], L. Girardello et al., Fujikawa[4], and Teshima[5] to the conclusion that SUSY at finite temperature is spontaneously broken. On the other hand, Van Hove[6] introduced an elegant averaging method called the graded-trace operation. When the order parameter is defined through the graded-trace method, it vanishes at finite temperature if SUSY is not broken at zero temperature. This fact led some people to state that SUSY is not broken at finite temperature if it is not broken at T=0. When TFD (for equilibrium) is applied to a model of SUSY, the Heisenberg equations are supersymmetric, while the thermal vacuum is not, because the thermally excited bosons and fermions have different distribution functions.[7] Thus in TFD the SUSY at finite temperature is spontaneously broken. These considerations indicate that the question of whether or not a thermally induced breaking of symmetry is spontaneous, is a linguistic matter; the answer depends on the formalism employed.

Putting aside this linguistic matter, we ask the question concerned with the Goldstone theorem, which states that a spontaneous breakdown of a continuous symmetry (or an appearance of an ordered state) needs the existence of a massless particle (or a quantum of long range correlation) called the Góldstone particle, unless the symmetry is a local gauge symmetry. This leads us to the question asking if a Goldstone fermion (Goldstino) appears in models of SUSY at finite temperature. This question was analysed by several people[4] by means of the Matsubara imaginary-time formalism. A complication in the application of the imaginary-time formalism to the study of SUSY at finite temperature arises from the need for an antiperiodic Grassmann parameter, which is introduced in order to compensate for the difference between the bosonic and fermionic boundary conditions associated with the imaginary-time interval. Surface-dependent terms appear from the integration boundary of the imaginary time in the Ward-Takahashi (WT) relations and make it difficult to obtain a clear answer to the question regarding the existence (or non-existence) of the Goldstone fermion. One the other hand, in ref. 8, the effective potential at finite temperature was analyzed by means of an analytic continuation in the complex time plane and the existence of the Goldstone fermion was suggested. However, the existence of the Goldstone fermion cannot be the general conclusion about the above question, because the thermally induced broken symmetries occur even when there is no interaction. It is obvious that a free particle model of SUSY without a massless fermion at T=0 cannot create a massless particle at T≠0 simply because the mass shift needs the presence of an interaction. We can therefore agree with the author of ref. 8 only when we interpret the conclusion of ref. 8 in the following way; the WT relations applied to SUSY at finite temperature indicate the presence of a zero-energy mode which is not necessarily a particle mode. (It cannot be a particle mode when there is no interaction.) Then our question is now what is the physical nature of this zero-energy mode? The obvious place to begin our analysis is the SUSY model without interaction. Use of TFD in this analysis led us to a very interesting result[7] and exposed a new question concerned with the interaction. The authors in ref. 9 presented a similar result for a free SUSY model.

A similar development is going on in study of the Lorentz symmetry at finite temperature. Recently, the author in ref. 10 concluded the presence of a Goldstone boson caused by the thermally induced breakdown of Lorentz symmetry. However, the generality of this conclusion is doubtful, because the thermally induced breakdown of Lorentz symmetry occurs even in a system of free fields which cannot create any new particles. We have recently analyzed[11] the thermally induced breakdown of Lorentz symmetry in a system of free fields. In the next section, we summarize our knowledge about the thermally induced broken supersymmetries for free fields and the results of our recent analysis on the effect of interactions.[12] In Sect. 3, the thermally induced broken symmetry of the Lorentz group is discussed in the free field model.

## 2. SUPERSYMMETRY

### Free Field Models

We now study a free field model of SUSY at finite temperature. We are interested in the question asking what kind of zero energy mode appears in the correlation function, which is the thermal vacuum expectation value (i.e. the thermal average) of the product of the generator of SUSY with some other operators. An easy intuition immediately leads us to an answer which is explained in the following. When the SUSY generator acts on the thermal vacuum, a thermally excited particle (either boson or fermion) changes its statistics, causing the reaction either boson $\rightarrow$ fermion or fermion $\rightarrow$ boson. These reactions carry no momentum transfer because the generator is a spatial integration of a density. Furthermore, since there is no interaction to create mass differences among the members of a supermultiplet, the above reactions do not need any energy transfer; they are zero-energy modes. These are the obvious candidates for the zero-energy mode which we are searching for.

In terminology of TFD, annihilation of a thermally excited particle is performed by a tilde-creation operator (say $\tilde{a}^{\dagger}(\beta)$) (here $\beta$ is referring to the inverse of temperature; $\beta = 1/k_B T$), while the usual creation process is taken care of by the non-tilde creation operator (say $a^{\dagger}(\beta)$). Thus, the above-mentioned reaction is caused by the pair operators, $a_B^\dagger(\beta)\tilde{a}_F^\dagger(\beta)$ or $a_F^\dagger(\beta)\tilde{a}_B^\dagger(\beta)$, where the suffix B(F) refers to boson (fermion). These pairs are called the "thermal superpairs".[7]

Let us now show how the picture for the usual Goldstone theorem is modified by a temperature effect.

Let us first remind readers of a brief and simple form of the proof of the usual Goldstone theorem. In general, the conserved charge of some symmetry of a Lagrangian is expressed in terms of the physical-particle operators (which are the asymptotic fields in the usual field theory without thermal freedom) as

$$Q = \sum_{i,j,\vec{p}} A_{ij} a_i^\dagger(\vec{p}) a_j(\vec{p}) + \sum_i (B_i \chi_i + B_i^* \chi_i^\dagger) \tag{2.1}$$

where $a_i$ stands for the annihilation operators of physical particles while $\chi_i$ are the annihilation operators of massless particles. The $B_i$'s are constant c-numbers. The time-independence of Q requires that the $\chi_i$ are massless and also that $A_{ij}$'s do not vanish only when the physical masses of i-th and j-th particles are degenerate. In the field theory without thermal freedom, we have $a_i^\dagger a_j|0\rangle = 0$. Since $Q|0\rangle \neq 0$ when the symmetry is spontaneously broken, the second term of (2.1) cannot vanish, implying the presence of massless particles. In TFD at finite temperature, however, the term $a_i^\dagger a_j$ fails to annihilate the thermal vacuum $|0(\beta)\rangle$ and, instead, we have $a_i^\dagger a_j|0(\beta)\rangle \propto a_i^\dagger \tilde{a}_j^\dagger|0(\beta)\rangle \neq 0$. Thus, the broken symmetry does not require the presence of the Goldstone massless particle ($\chi_i$) and the thermal pair $a_i^\dagger \tilde{a}_j^\dagger$ acts as a non-trivial zero-energy mode, because $\tilde{a}_j^\dagger$ creates a negative energy. (i.e. $\tilde{a}_j^\dagger$ annihilates a thermally excited particle.) We should, however, further elaborate the above argument, because the symmetry generator in TFD is not Q but $\hat{Q} \equiv Q - \tilde{Q}$, where $\tilde{Q}$ has a structure similar to Q, in which the non-tilde operators are replaced by tilde operators. Thus, the condition for symmetry breakdown in TFD has the form: $\hat{Q}|0(\beta)\rangle \neq 0$. It can be shown[12] that, when $Q$ is bosonic, the contribution of thermal pairs to $Q|0(\beta)\rangle$ is the same as the one to $\tilde{Q}|0(\beta)\rangle$ and, therefore, these contributions compensate each other, leading to a vanishing contribution of thermal pairs in $\hat{Q}|0(\beta)\rangle$. Then, the presence of the massless field $\chi_i$ is again demanded. However, when $Q$ is fermionic, because of the difference of the distribution functions between fermions and bosons, the effect of thermal pairs in $\hat{Q}|0(\beta)\rangle$ does not cancel.[12] In this case the presence of massless fermions is not always demanded and the thermal superpair provides the zero-energy mode. This is the situation of free SUSY model at finite temperature.

In order to elucidate the contribution of thermal superpairs to the correlation functions consisting of $\hat{Q}$ and another local operator, we calculate a simple example of a loop consisting of a fermion line and a boson line, because the above-mentioned correlation function for SUSY without interaction is the zero-momentum part of such a loop. As was shown by Eq. (4.1; 122) in chapter 4 of the book in ref. 3,

$$\int \frac{d^3\ell}{(2\pi)^3} \int \frac{d\ell_o}{2\pi} \, [U(\omega_+)\{k_o-\ell_o-\omega_+ + i\epsilon\tau\}^{-1}U^\dagger(\omega_+)]_{\alpha\beta}$$

$$\times \, [U_B(\omega_-)\tau\{\ell_o^2-(\omega_- - i\epsilon\tau)^2\}^{-1}U_B(\omega_-)]_{\alpha\beta}$$

$$= i\int_{-\infty}^{\infty} d\kappa \; \sigma(\kappa;\omega_+,\omega_-)\,[U(\kappa)\{k_o-\kappa+i\epsilon\tau\}^{-1}U^\dagger(\kappa)]_{\alpha\beta} \qquad (2.2)$$

where $\omega_+ = \{(\vec{k}+\vec{\ell})^2 + m^2\}^{1/2}$ and $\omega_- = (\vec{\ell}^2 + m^2)^{1/2}$. Since there is no interaction to create a boson-fermion mass difference, the masses of bosons and fermions are degenerate; they are denoted by $m$. The matrices, $U(\omega)$ and $U_B(\omega)$, are the thermal matrices for fermions and bosons, respectively. Since the explicit structure of $U(\omega)$ and $U_B(\omega)$ is not essential in the following argument and their form can be found in many references (e.g. (4.1.76) and (4.1.86) in the book in ref. 3), we do not write them here.

The essential point is the structure of the spectral function $\sigma(\kappa;\omega_+,\omega_-)$:

$$\sigma(\kappa;\omega_+,\omega_-) = -\frac{1}{2\omega_-}[\delta(\kappa-\omega_+-\omega_-)\{f_B(\omega_-)+1-f_F(\omega_+)\}$$

$$-\delta(\kappa-\omega_++\omega_-)\{f_B(-\omega_-)+1-f_F(\omega_+)\}], \qquad (2.3a)$$

where $f_B$ and $f_F$ are the distribution functions of bosons and fermions, respectively:

$$f_B(\omega) = \frac{1}{e^{\beta\omega}-1},$$

$$f_F(\omega) = \frac{1}{e^{\beta\omega}+1}.$$

Since with $\vec{k}=0$, $\omega_+ = \omega_- = (\vec{\ell}^2+m^2)^{1/2}$ and since $f_B(\omega)+1 = -f_B(-\omega)$, the second term in (2.3) is the contribution of thermal superpair which with $\vec{k}=0$ creates the $\delta(\kappa)$-term as

$$\delta(\kappa)\{f_B(\omega_\ell) + f_F(\omega_\ell)\}.$$

This is the zero-energy singularity in the fermionic correlation function under consideration.

It is instructive to compare the above result with a loop consisting of two fermion lines or of two boson lines. The spectral function of a two fermion loop (cf. Eq. (4.1.121) in ref. 3) is

$$\sigma(\kappa;\omega_+,\omega_-) = \delta(\kappa-\omega_++\omega_-)[f_F(\omega_+)-f_F(\omega_-)], \qquad (2.3b)$$

while the contribution of thermal boson pairs in the spectral function of a two boson loop is

$$\sigma(\kappa;\omega_+,\omega_-) = \delta(\kappa-\omega_++\omega_-)[f_B(\omega_+)-f_B(\omega_-)]. \qquad (2.3c)$$

These spectral functions vanish at $\vec{k}=0$, and therefore there is no effect of thermal pairs. This is intuitively understandable, because the creation of these thermal pairs corresponds to reactions in which no thermally excited particles change their states, contrary to the case of the creation of thermal superpairs, which is the reaction in which a thermally excited particle changes its statistics. These results are in conformity with the previous considerations about (2.1).

We now illustrate our consideration by use of the Wess-Zumino model.[13] The Lagrangian density of this model in TFD[7,12] is

$$\hat{\mathcal{L}} = \sum_{\alpha=1}^{2} [-\frac{1}{2}\varepsilon^{\alpha}\partial_{\mu}A^{\alpha}\partial^{\mu}A^{\alpha} - \frac{1}{2}\varepsilon^{\alpha}\partial_{\mu}B^{\alpha}\partial^{\mu}B^{\alpha}$$

$$-\frac{i}{2}\bar{\psi}^{\alpha}\not{\partial}\psi^{\alpha} + \frac{\varepsilon^{\alpha}}{2}(F^{\alpha})^{2} + \frac{\varepsilon^{\alpha}}{2}(G^{\alpha})^{2}$$

$$+ m\,\varepsilon^{\alpha}(F^{\alpha}A^{\alpha}+G^{\alpha}B^{\alpha} - \frac{i}{2}\varepsilon^{\alpha}\bar{\psi}^{\alpha}\psi^{\alpha})$$

$$+\frac{g}{2}\varepsilon^{\alpha}\{F^{\alpha}(A^{\alpha 2}-B^{\alpha 2}) + 2G^{\alpha}A^{\alpha}B^{\alpha}$$

$$- i\varepsilon^{\alpha}\bar{\psi}^{\alpha}(A^{\alpha}-\gamma_{5}B^{\alpha})\psi^{\alpha}\}] \tag{2.4}$$

where $\varepsilon^{1}=1$, $\varepsilon^{2}=-1$. We used the thermal doublet notation:

$$A = \begin{pmatrix} A \\ \tilde{A} \end{pmatrix},\ B = \begin{pmatrix} B \\ \tilde{B} \end{pmatrix},\ \psi^{\alpha} = \begin{pmatrix} \psi \\ \tilde{\psi} \end{pmatrix},\ F^{\alpha} = \begin{pmatrix} F \\ \tilde{F} \end{pmatrix} \text{ and } G = \begin{pmatrix} G \\ \tilde{G} \end{pmatrix}. \tag{2.5}$$

$\psi$ is the fermion while A and B are real scalar fields. The real scalars F and G are the auxiliary fields. This model is invariant under the SUSY transformation which is generated by the supercharge $Q^{\alpha}$ of the form[12].

$$Q^{\alpha} = \int d^{3}x \ \mathcal{J}^{\alpha 0} \tag{2.6}$$

$$\mathcal{J}^{\alpha}_{\mu} \equiv [\not{\partial}(A^{\alpha}-\gamma_{5}B^{\alpha}) + m(A^{\alpha}+\gamma_{5}B^{\alpha})$$

$$+\frac{g}{2}(A^{\alpha}+\gamma_{5}B^{\alpha})^{2}]\gamma_{\mu}\psi^{\alpha} \tag{2.7}$$

To discuss the WT relation in a simplified form, let us consider

$$\int d^{3}x <0(\beta)|[\mathcal{J}^{\alpha 0}(x), M^{\beta}(y)]_{+}|0(\beta)>$$

$$= \int dq^{0}\, e^{iq^{0}(x^{0}-y^{0})}\sigma_{M}(q^{0},\vec{0})\,\delta^{\alpha\beta} \tag{2.8}$$

Here $M^{1}(y)$ is a local fermionic operator M(y) which consists only of non-tilde operators, while $M^{2}(y)=\tilde{M}^{\dagger}(y)$. Since the supercharge is independent of time, we have

$$q^{0}\sigma_{M}(q^{0},0) = 0, \tag{2.9}$$

which gives

$$\sigma_M(q^o,0) = a_M\delta(q^o) \tag{2.10}$$

with a constant $a_M$. It can be shown that, when $M = \bar{\psi}$, we have

$$a_{\bar{\psi}} = f \equiv \langle 0(\beta)|F^1(x)|0(\beta)\rangle$$

$$= \langle 0(\beta)|F^2(x)|0(\beta)\rangle. \tag{2.11}$$

We have also that, when $M = A\bar{\psi}$,

$$a_{A\bar{\psi}} \neq 0. \tag{2.12}$$

This shows that there should be a zero-energy mode in the $A\bar{\psi}$-channel.

Let us now consider the case without interaction (g=0). Then, we have $a_{\bar{\psi}}=0$, implying that there is no zero-energy mode in the $\psi$-channel. On the other hand, (2.8) with $M = A\bar{\psi}$ is the $\delta$-function part of a fermion and boson loop. Thus, as the previous loop calculation indicated, the zero-energy mode in the $A\bar{\psi}$-channel is the thermal superpair.

Interacting Model of SUSY

We now consider the Wess-Zumino model with an interaction at finite temperature.[12] It is assumed that SUSY is not broken at $T = 0$.

Since the WT relations cannot tell us the nature of the zero energy mode in the present situation, we must perform certain dynamical calculations.

When $T \neq 0$, we have

$$v = \langle 0(\beta)A^\alpha|0(\beta)\rangle \neq 0, \tag{2.13}$$

$$f = \langle 0(\beta)F^\alpha|0(\beta)\rangle \neq 0. \tag{2.14}$$

If non-vanishing v and f would appear in the tree approximation, we could rely on the usual loop expansion method in the analysis of WT relations. In the case under consideration, nonvanishing v and f are loop contributions, and we are forced to modify the loop expansion method. Even when we consider only the one loop contributions to v and f, any one loop consisting of Feynman lines with renormalized mass is not really one loop, because the renormalized mass containing v and f includes already some loop corrections. In order to achieve a self-consistency in the analysis of the WT relations, we need to assemble an infinite number of diagrams. The rule for choosing these diagrams in the one-loop approximation was given in ref. 12: we first evaluate the one-particle irreducible N-point vertices in the usual one-loop approximation as the "building blocks", then we sum up all possible tree diagrams constructed out of these N-point vertices ($N \geq 3$) and propagators whose inverse is evaluated in one-loop level. This is called the modified one-loop approximation.[12] Its extension to the modified higher-loop approximation was also suggested in ref. 12. It was shown by explicit computations that the WT relations are satisfied by the modified one-loop approximation.

Leaving details to ref. 12, let us briefly summarize the physical picture which emerges from the calculations. The nonvanishing v and f create a mass difference between boson and fermion. Similarly to the free field case, $a_{A\bar{\psi}} \neq 0$ indicates that a zero-energy mode appears in the $A\bar{\psi}$-channel. Furthermore, since the interaction makes $a_{\bar{\psi}} = f \neq 0$, the zero-energy mode appears also in the $\bar{\psi}$-channel. (This was first pointed out in ref. 8.) This can be explicitly confirmed by the modified one-loop calculation. The modified one-loop calculation also shows that there remains a massive mode in the $\bar{\psi}$-channel. In other words, the zero energy mode appears in a combination $A\bar{\psi} - \alpha\bar{\psi}$, while the massive mode appears in $\bar{\psi} - \beta A\bar{\psi}$ where the coefficients $\alpha$ and $\beta$ vanish at $g = 0$. This indicates that the massive mode is the original massive fermion modified by the self-energy. In the limit $g \to 0$, the zero-energy mode becomes the thermal superpair as we have seen in the previous section, suggesting that the zero-energy mode is a fermionic bound state consisting of a thermal state in which the binding energy is compensated by the mass difference. Intuitively speaking, this zero-energy mode is a transmutation of the statistics of a thermally excited particle (as it was in the case of free field). Through the interaction effect this transmutation induces certain modifications among the particles, leading to a change of energy which compensates the mass difference. This interaction effect is represented by the infinite chain diagrams in the modified one-loop approximation. With reasonable computational results together with the intuitive picture, we feel that the problem of SUSY at finite temperature is now well understood. In order to further confirm our view, we are, at present, involved[14] in the O'Raifeartaigh model[15] which has a very rich symmetry content.

## 3. THE LORENTZ SYMMETRY AT FINITE TEMPERATURE

In this section we present an account of our recent study of Lorentz symmetry at finite temperature. To save space, we show here only a brief sketch and a detailed consideration will be given in a separate paper.[11]

Let us first list some relations which are of relevance to our study of the Lorentz symmetry at finite temperature. Using TFD, the Lorentz transformation generator has the form $\hat{L}_{\mu\nu} = L_{\mu\nu} - \tilde{L}_{\mu\nu}$. Denoting the energy stress tensor by $T_{\mu\nu}$, we have

$$
\begin{aligned}
[\hat{L}_{\mu\nu}, T_{\lambda\rho}] &= -i(x_\mu \partial_\nu - \chi_\nu \partial_\mu) T_{\lambda\rho}(x) \\
&\quad + i(g_{\mu\lambda} T_{\nu\rho}(x) - g_{\nu\lambda} T_{\mu\rho}(x) \\
&\quad + g_{\mu\rho} T_{\lambda\nu}(x) - g_{\nu\rho} T_{\lambda\mu}(x)). \qquad (3.1)
\end{aligned}
$$

At T=0, we have $\langle 0|T_{\mu\nu}|0\rangle \propto g_{\mu\nu}$ which gives $\langle 0|[\hat{L}_{\mu\nu}, T_{\lambda\rho}]|0\rangle = 0$. However, with $T \neq 0$, we have

$$
\langle 0(\beta)|T_{\mu\nu}(x)|0(\beta)\rangle = g_{\mu\nu} I_o(T) + g^o_\mu g^o_\nu I_1(T) \quad \text{with } I_1(T) \neq 0, \qquad (3.2)
$$

since the contribution from the thermal distribution distinguishes the time components from the space components. This gives

$$\langle 0(\beta)|[\hat{L}_{\mu\nu},T_{\lambda\rho}(x)]|0(\beta)\rangle = i(g_{\mu\lambda}\overset{o}{g}_{\nu}\overset{o}{g}_{\rho} - g_{\nu\lambda}\overset{o}{g}_{\mu}\overset{o}{g}_{\rho} + g_{\mu\rho}\overset{o}{g}_{\lambda}\overset{o}{g}_{\nu} - g_{\nu\rho}\overset{o}{g}_{\lambda}\overset{o}{g}_{\mu})I_1(T). \tag{3.3}$$

The non-vanishing components are, in short, given by

$$\langle 0(\beta)|[\hat{L}_{oi},T_{oj}(x)]|0(\beta)\rangle = (-i)g_{ij}I_1(T) \neq 0. \tag{3.4}$$

which leads to

$$\hat{L}_{oi}|0(\beta)\rangle \neq 0. \tag{3.5}$$

Though the Heisenberg equation is Lorentz invariant, (3.5) implies that the thermal vacuum is not Lorentz invariant. Thus, according to the terminology of TFD, the Lorentz symmetry is spontaneously broken. Furthermore, $I_1(T)\neq 0$ implies that there is a zero-frequency singularity in the correlation function consisting of $L_{oi}$ and $T_{oj}$. (Using a model of a real scalar field $\phi$ and going through a different route, the author of ref. 10 recently concluded that the correlation function consisting of $L_{oi}$ and $\phi(x)\phi(y)$ has a zero-energy singularity.) However one cannot immediately conclude the presence of a massless particle from (3.4) because this is true even when there is no interaction to create a massless particle. In the case of free field models, the action of $\hat{L}_{\mu\nu}$ on the thermal vacuum $|0(\beta)\rangle$ is only to boost each of the thermally excited particles. Therefore, no obvious zero-energy mode exists except the thermal pair of the $a_B^{\dagger}(\beta)\tilde{a}_B^{\dagger}(\beta)$-type with vanishing total momentum. However, as was shown by (2.3c) in section 2, its contribution to the correlation function of the bosonic operators seems to vanish, manifesting the fact that the creation of thermal pair with zero momentum is a trivial process; every thermally excited boson stays as it is. That is, although there exists a candidate for the zero energy mode consisting of the thermal pair, such contribution seems to vanish according to the general arguments in Sect. 2.

To escape the paradox is to recall that the spectral representation of correlation functions can tell us about the physical modes when and only when the operators in the correlation function do not carry explicit t or $\vec{x}$. Since $\hat{L}_{oi}$ contains t and $x_i$, we need to carefully analyze the structure of the correlation functions in order to identify the cause of the zero-frequency singularity mentioned above.

To proceed further, we consider a free scalar field $\phi(x)$. Then we find

$$I_1(T) = \frac{1}{(2\pi)^3} \int d^3k \, \frac{1}{3\omega_k} \, \frac{1}{e^{\beta\omega_k}-1} \, (4\omega_k^2 - m^2) \tag{3.6}$$

and

$$\hat{L}_{oi} = t\hat{p}_i - \int d^3x \, x_i \hat{H}(x) \tag{3.7}$$

where $\hat{P}_i = P_i - \tilde{P}_i$ with $\vec{P}$ being the momentum operator, while $\hat{H}(x) = H(x) - \tilde{H}(x)$ with $H(x)$ being the energy density $T_{oo}(x)$. Since $\hat{P}_i|0(\beta)\rangle = 0$, we have

$$\langle 0(\beta)|[\hat{L}_{oi}, T_{oj}(y)]|0(\beta)\rangle$$
$$= -\int d^3x \, x_i \langle 0(\beta)|[H(x), T_{oj}(y)]|0(\beta)\rangle. \tag{3.8}$$

Since neither $H(x)$ nor $T_{oj}(y)$ contain t or $\vec{x}$ explicitly, we can write down the spectral representation:

$$\langle 0(\beta)|[H(x), T_{oj}(y)]|0(\beta)\rangle$$
$$= \frac{1}{(2\pi)^3} \int d^4p \, e^{-ip(x-y)} \sigma_j(p). \tag{3.9}$$

We see from (2.3c) that $\sigma(p)$ contains a term of the form

$$\delta(p_o - \omega_+ + \omega_-)[f_B(\omega_+) - f_B(\omega_-)] \tag{3.10}$$

with $\omega_+ = ((\vec{p}+\vec{\ell})^2 + m^2)^{1/2}$ and $\omega_- = (\vec{\ell}^2 + m^2)^{1/2}$.

On the other hand, since $x_i$ in (3.8) acts as $\partial/\partial p_i$ we have

$$<0(\beta)|[\hat{L}_{oi}, T_{oj}(y)]|0(\beta)> = i\int dp_o [\frac{\partial}{\partial p_i}\sigma_j(p)]_{\vec{p}=0} \qquad (3.11)$$

According to (3.11), $(\partial/\partial p_i)\sigma_j(p)$ with $\vec{p}=0$ contains a term which is proportional to $\delta(p_o)[\partial f_B(\omega)/\partial\omega]$ with $\omega=\omega_\ell$. Therefore, although $\sigma_j(p)$ with $\vec{p}=0$ has no term with a $\delta(p_o)$-singularity, $(\partial/\partial p_i)\sigma_j(p)$ with $\vec{p}=0$ does have a term with a $\delta(p_o)$-singularity.

The moral we learn from the above consideration is the following. Correlation functions among bosonic operators (which do not explicitly contain either t or $\vec{x}$), never carry the thermal pair zero-energy mode. However, when an operator contains t or $\vec{x}$ explicitly, the thermal pair acquires an infinitesimal energy or momentum, and therefore, becomes visible in the correlation functions of such operators. The thermal pair with an infinitesimal momentum (or energy) means that a thermally excited particle changes its momentum or energy by an infinitesimal but not zero amount.

It seems quite likely that this situation will remain, even when there are interactions, since even with the interactions, each member of the thermal pair has the same energy.

Most of the above results for the Lorentz symmetry at finite temperature can be applied also to the symmetry of the Galilei transformation at finite temperature.[11] Furthermore, the Galilei invariance can be spontaneously broken even in the free model at zero temperature, when one considers a free fermi gas filled up to the fermi energy.

## ACKNOWLEDGEMENT

The authors would like to thank Drs. N.J. Papastamatiou and M. Nakahara for helpful discussions, and Dr. H. Aoyama for sending us his work prior to its publication. This work was supported by the Natural Science and Engineering Research Council, Canada, and by the Dean of the Faculty of Science, The University of Alberta, Edmonton, Alberta, Canada.

References

1. I. Ojima, in "Lecture Notes in Physics No. 176, Gauge Theory and Gravitation, Proceedings, Nara, Japan 1982" (Springer).
2. A. Das and M. Kaku, Phys. Rev. D18, 4540 (1978).
3. H. Umezawa, H. Matsumoto and M. Tachiki, "Thermo Field Dynamics and Condensed States", (North-Holland Pub., Amsterdam, New York, Oxford, 1982).
4. L. Girardello, M. Grisaru and P. Salomonson, Nucl. Phys. B178, 331 (1981); K. Fujikawa, Z. Phys. C - Particles and Fields 15, 275 (1982).
5. K. Teshima, Phys. Letters 123B, 226 (1983).
6. L. Van Hove, Nucl. Phys. B207, 15 (1982).
7. H. Matsumoto, N. Nakahara, Y. Nakano and H. Umezawa, Phys. Letters 140B, 53 (1984); Phys. Rev. D29, 2838 (1984).
8. D. Boyanovsky, Phys. Rev. D29, 743 (1984).
9. H. Aoyama and D. Boyanovsky, Phys. Rev. D30, 1356 (1984).
10. H. Aoyama, (Preprint, SLAC-PUB-3723, June, 1985).
11. H. Matsumoto, N.J. Papastamatiou, H. Umezawa and N. Yamamoto, (in preparation).
12. H. Matsumoto, M. Nakahara, H. Umezawa and N. Yamamoto, (preprint, Univ. of Alberta, 1985).
13. J. Wess and B. Zumino, Nucl. Phys. B70, 39 (1974).
14. Yvan Leblanc and H. Umezawa, (in preparation).
15. L. O'Raifeartaigh, Nucl. Phys. B96, 331 (1975).

# RING DIAGRAMS AND PHASE TRANSITIONS

Koichi Takahashi
Department of Physics, College of General Education
Tohoku Gakuin University
Sendai 980, Japan

ABSTRACT

Ring diagrams at finite temperatures carry most infrared-singular parts among Feynman diagrams. Their effect to effective potentials are in general so significant that one must incorporate them as well as 1-loop diagrams. We see these circumstances in some examples of supercooled phase transitions.

## 1. INTRODUCTION

The natures of finite temperature systems of elementary particles have been studied so far by evaluating 1-loop effective potentials. They provide us informations about phase transitions which affected the way of the evolution of our universe.

Higher-order corrections start by taking ring diagrams into account/1/. The ring diagrams are those whose skeletons are of

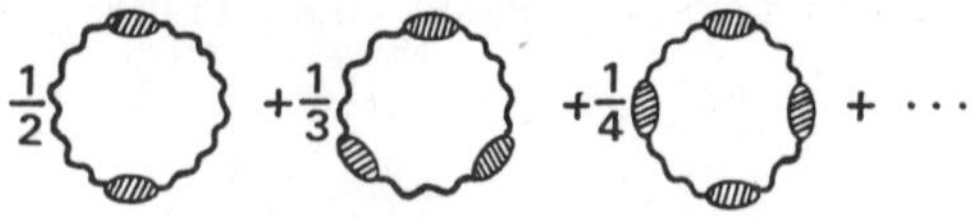

Fig.1 Ring diagrams

1-loop, i.e. 1-loop diagrams with self-energy insertions (see Fig.1). This type of diagrams has already been studied in the earlier work of Dolan and Jackiw/2/ (they called them 'daisy' diagrams for $\phi^4$ theory) when, together with other authors, they developed the relativistic finite temperature field theories. However, the potential importance of ring diagrams in effective potentials have never been noticed so far.

In massless theories, for each order of coupling constant g, the ring diagrams are the most infrared singular ones. The infrared singularities relevant to the discussions presented here owe their origin to the bosonic zero-energy modes which are peculiar to finite temperature field theories with imaginary time (i.e. Matsubara) formalism (Although the real time formalism as the thermo field dynamics (see the contribution by Umezawa, Matsumoto and Yamamoto in this book) provides us physically transparent pictures of interacting particles, we here use the imaginary formalism for calculational convenience for infrared singularities). The leading 1-loop contribution is of order $g^2$,while the exclusive sum of ring diagrams gives finite result of order $g^3$. In the followings, we shall see the importance of these contributions in concrete examples of a scalar and gauge models.

# 2. A TOY MODEL

## 2.1 Effective Potential and Bounce Action

The effect of ring diagrams at finite temperature T can easily be seen in the massless scalar theory with interaction $(g^2/4)\phi^4$ * Let us consider the dimensionless function $f(x)$($x=g\phi/T$, $\phi$ being the classical field):

$$f(x) \equiv V_{eff}/NT^4 = (V_1 + V_R)/NT^4 = f_1(x) + f_R(x), \tag{1a}$$

$$f_1(x) = \frac{1}{32\pi^2}x^4(\ln x - \ln x_0 - \frac{1}{2}) + \frac{1}{2\pi^2}I(x), \quad x_0 = M/T, \tag{1b}$$

$$f_R(x) = -\frac{1}{12\pi}((p + x^2)^{3/2} - x^3 - p^{3/2}), \quad p=\frac{Ng^2}{12}, \tag{1c}$$

$$I(x) = \int_0^\infty y^2 dy \ln(1 - \exp-(y^2 + x^2)^{1/2}) + 2.165... \tag{1d}$$

In the above expression, $f_1(x)$ and $f_R(x)$ correspond to 1-loop and ring diagrams, respectively and are adjusted so that $f_1(0)=f_R(0)=0$ by adding some constants. p in (1c) comes from the vacuum polarization and, roughly speaking, the vacuum polarization is proportional to the number of particle species (it is exact in our toy model). A large particle number produces a large p.

One can expand f(x) around x=0. The coefficient of $x^2$ term is $1/24-\sqrt{p}/8\pi$ Then, one has a condition for the symmetry restoration at high temperatures

---

* There are some arguments/3/ that $\phi^4$ theories may be trivial in 4-dimension. However, we concern ourselves with their perturbative natures here.

$$Ng^2 < \frac{4\pi^2}{3} \tag{2}$$

The significance of the ring diagram effect to the super-cooled phase transition can be seen by comparing tunneling probabilities for $f_1(x)$ and $f(x)$. In the lowest order of perturbations, they are determined by 3-dimensional bounce actions. For their estimations, we may use the formula

$$A = c\frac{D^3}{\sqrt{H}}, \tag{3}$$

where D is the zero-point of the effective potential, and H is the height of the barrier. c is a numerical constant. With c=3.3405..., eq.(3) is exact for potentials with quadratic and quartic terms only*.(When this formula is applied to scaled effective potentials ($f(x)$ or $f_1(x)$), then a factor $1/\sqrt{N}\,g^3$ must be multiplied to the above A to obtain the true action. Furthermore, in actual models studied in following chapters, g

* It is easy to show that, when the potential is scaled as $V \to aV$, the 3-dimensional action is scaled as $A \to A/\sqrt{a}$ . So, the characteristic dependence of the bounce action on the height of the barrier H will be $1/\sqrt{H}$ . If we choose D as the other quantity which characterizes the value of action, then the form of Eq.(3) is unique apart from the numerical coefficient. It will turn out that appropriate values of c are 4.1 for 1-loop potential and 3.8 for 1-loop+ring potential in the temperature region considered.

has a non-negligible temperature dependences. In this chapter, we neglect it.)

In Table.1, values of scaled bounce actions for various p and T are given (Multiplications of $1/\sqrt{N}\, g^3$ yield true actions). We see 25% reduction of action for e.g. p=0.4 and M/T=500. If one assumes the form $1/\ln x = 1/\ln(M/T)$ for the temperature dependence of bounce action for $f_1$/4/, further lowering of the temperature as $M/T = (500)^{4/3} = 4000$ will be required in order to attain such a small value of action. This means that, in

| p | 0 | 0.1 | 0.2 | 0.3 | 0.4 | 0.5 |
|---|---|---|---|---|---|---|
| A | 54 | 50 | 46 | 43 | 40 | 36 |
| $3.8D^3/\sqrt{H}$ | 50 | 49 | 45 | 43 | 40 | 37 |

Table 1.a Computer calculations of p dependence of bounce actions at M/T=500. p=0 corresponds to potentials without ring diagram corrections. Formula (3) reproduces the actions very nicely.

| M/T | 10 | $10^2$ | $10^3$ | $10^4$ | $10^5$ | $10^6$ |
|---|---|---|---|---|---|---|
| 1-loop | 190 | 75 | 49 | 37 | 30 | 25 |
| p=0.3 | 140 | 59 | 39 | 29 | 23 | 20 |

Table 1.b Temperature dependence of bounce actions.

actual phase transitions during the evolution of the universe, entropy productions might be suppressed to quantities much smaller than that of the 1-loop estimations.

## 2.2 Entropy Production

The rapid expansion of the universe yieldes a large amount of entropy at the supercooled phase transitions. In the real world, this has an effect to reduce the baryon number to entropy ratio generated after the GUT phase transition. In our toy model, qualitative estimations of entropy productions for various p can easily be performed by following previous works/4,5/.

The expansion time scale of the universe in the de'Sitter phase is $M_P/\sqrt{\rho}$, where $M_P$ is the Planck mass and $\rho$ is the energy density of the false vacuum. For our toy model, $\rho = \frac{NM^4}{128\pi^2}$. Then, the transition will occur when the transition probability per unit 4-space multiplied by the causal 4-space volume gets nearly equal to one:

$$\frac{M_P}{\sqrt{\rho}} M^4 \exp(-A(T^*)) \cong 1, \text{ or } \quad A(T^*) = 4\ln\frac{M_P}{M} + C. \tag{4}$$

C may be a numerical constant whose value is expected much smaller than the characteristic values of A(T*) etc. We set it to be 0 for simplicity in our calculations.

The temperature $T_a$ after the transition is given roughly by $T_a = (\rho/N)^{1/4}$. Then, the ratio of the final entropy to the initial one is

$$\frac{S_f}{S_i} \cong \left(\frac{T_a}{T^*}\right) \cong \left[\frac{1}{(128\pi^2)^{1/4}}\frac{M}{T^*}\right]^3 \tag{5}$$

The results of numerical calculations are summarized in Table 2. (Absolute values should not be taken too seriously. Important are the differences between results of 1-loop and 1-loop+ring calculations.) We see that, for our effective potential

| $M_p/M$; N | $10^5$ ; 30 | | $10^9$ ; 20 | | $10^{17}$ ; 10 | |
|---|---|---|---|---|---|---|
| p | 1-loop | 0.3 | 1-loop | 0.3 | 1-loop | 0.3 |
| T*(GeV) | 1 | $5\times10^2$ | $5\times10^4$ | $1\times10^6$ | 3 | 50 |
| $S_f/S_i$ | $5\times10^{39}$ | $4\times10^{31}$ | $4\times10^{13}$ | $5\times10^9$ | 100 | 40 |

Table 2 Transitions temperatures and entropy productions. ($M_P=10^{19}$GeV)

f(x), the transition temperature T* is high and the increase of entropy is small. For p not to be too small, the ring diagram effects should not be neglected.

## 3. UNIFIED THEORIES

The extension of the previous calculations to unified models is straightforward. Here, we give arguments for theories with the symmetry of 1.SU(5) and 2.SU(2)×U(1).

### 3.1 SU(5) Theory

We are not interested in the entropy produced, since it is not relevant to baryon number to entropy ratio which is produced after the breaking of SU(5). So, we give only the transition temperature T*. Two directions of the symmetry breaking are considered ($R_\xi$ gauge with $\xi = \infty$ is adopted):

i. SU(4)×U(1): $\mathrm{diag}(\phi_{adj})=\sqrt{1/20}\,\phi(1,1,1,1,-4)$

$$V_R = -8g^3\left[\left(\frac{C}{9}T^2 + \frac{5}{8}\phi^2\right)^{3/2} - \left(\frac{C}{9}\right)^{3/2} T^3\right]\cdot\frac{T}{4\pi}$$

ii. SU(3)×SU(2)×U(1): $\mathrm{diag}(\phi_{adj})=\sqrt{2/15}\,\phi(1,1,1,-3/2,-3/2)$

$$V_R = -12g^3\left[\left(\frac{C}{9}T^2 + \frac{5}{12}\phi^2\right)^{3/2} - \left(\frac{C}{9}\right)^{3/2} T^3\right]\cdot\frac{T}{4\pi}$$

Here, $C=5+N_g+\frac{1}{2}(N_H+5N_A)$, $N_g$, $N_H$ and $N_A$ being the number of generation, $\underline{5}$Higgs and $\underline{24}$Higgs multiplets, respectively. We set $N_g=3$, $N_H=N_A=1$ in calculations(the contribution of Higgs field is neglected).

As the temperature dependence of g, we choose the following form:

$$\frac{1}{g^2} = \frac{1}{g_0^2} + \frac{1}{24\pi^2}\left(55-4N_g-\frac{1}{2}(N_H+5N_A)\right)\ln\frac{M}{T}, \quad M=5\times10^{14}\,\text{GeV}.$$

The above $V_R$'s are added to the 1-loop effective potentials

$$V_1 = \frac{3}{64\pi^2}C(Dx^2)^2\left(\ln(Dx^2)-\ln\left(\frac{M}{T}\right)^2-\frac{1}{2}\right) + \frac{1}{2\pi^2}CI(\sqrt{Dx}),$$

$$(C,D) = \left(8,\frac{5}{8}\right) \text{ for } SU(4)\times U(1);\ \left(12,\frac{5}{12}\right) \text{ for } SU(3)\times SU(2)\times U(1)$$

to obtain total effective potentials for our SU(5) model(we omit $\phi^4$ self couplings of Higgs fields in order to see pure ring diagram effects). The symmetry restoration condition which is an analogue of eq.(2) is satisfied if g is asymptotically free.

At some temperatures, we calculate effective potentials numerically and determine A by using the formula (3) with c=4.1 for 1-loop and c=3.8 for 1-loop+ring potentials. Then, eq.(4) is solved. The transition will be induced when the action takes a value of 40. The results are given in Table 2. We see that ring diagrams give a striking effect to raise the transition temperature T* by one order of magnitude. This corresponds to the reduction of entropy production by three order of magnitude.

| | SU(4)× U(1) | | SU(3)× SU(2)× U(1) | |
|---|---|---|---|---|
| potential | 1-loop | 1-loop+ring | 1-loop | 1-loop+ring |
| T*(GeV) | $2\times10^{10}$ | $3\times10^{11}$ | $4\times10^{9}$ | $6\times10^{10}$ |

Table 3 Transitions temperatures for SU(5) theory

3.2 SU(2)×U(1) Theory In A Nutshell

It seems necessary to invoke the chiral symmetry breaking to avoid the extreme supercooling which causes too much entropy production at the stage of SU(2)×U(1) breaking in the universe evolution/4/. Then, the transition temperature T* is expected to be of order 200MeV. However, taking the ring diagram effect from the pure SU(2)×U(1) force into account, we obtain a similar result for T* of the full gauge symmetry breaking. See Fig.2. The detailes will be reported elsewhere/6/.

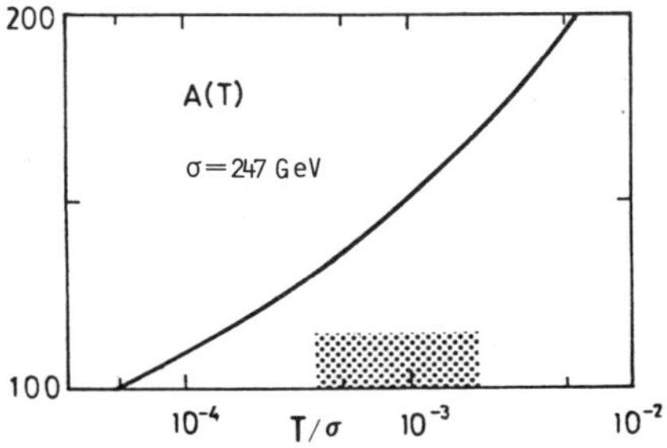


Fig.2 The temperature dependence of bounce action with the ring diagram correction. The transition to U(1) vacuum will occur in the shaded region of temperature.

## 4. CONCLUSION

The ring diagrams in massless theories have in general important effects in the supercooled phase transitions. This is due to the largeness of numbers of massless particles before the symmetry breaks. Another important cause is the temperature dependences of coupling constants. Although these have been couched in some specific models, their generality in models of Coleman-Weinberg type seems almost apparent. When phase transitions are investigated by the perturbative method, we will always have to pay enough attention to ring diagrams.

## REFERENCES


1 B.A.Freedman and L.D.McLerran: Phys.Rev.D16(1977),1130 1146
J.I.Kapusta: Nucl.Phys.B148(1979),461
K.Takahashi: Z.Phys.-C 26(1985),601

2 L.Dolan and R.Jackiw: Phys.Rev.D9(1974),3320

3 C.A. de Carvalho, C.S.Carracciolo and J.Frohlich: Nucl.Phys. B215(FS7)(1983),209 ;and references therein.

4 E.Witten: Nucl.Phys.B177(1980),477

5 S.Weinberg: 'Gravitation and Cosmology',Wiley(1972)

6 K.Takahashi: To be published in Phys.Rev.Lett.(the issue of 6 January 1986)

# Scalar Confinement Potential in QCD

H.Suura
School of Physics and Astronomy, University of Minnesota,
Minneapolis, Minn. 55455

A wave equation in the confining range is derived in QCD based on a linear flux approximation and a large N approximation. It shows existence of a Lorentz vector (energy-type) as well as a Lorentz scalar (mass-type) confinement potentials in general, but no scalar potential with the Lorentz structure corresponding to a scalar-boson exchange. If the vector confinement potential should be small, there will be no spin-orbit term from the long range potential.

It has been claimed[1)] that a Lorentz-scalar confinement potential is definitely favored over a vector-type to account for the fine structure of heavy quarkonia. In most of literature the scalar- and the vector-type potentials are defined as having Dirac-matrix structures corresponding to one scalar or vector boson exchange. Thus, one would write a Breit-type equation.

$$E\chi(1,2)=(-i\alpha\cdot\nabla_1+m_1\beta)\chi(1,2)$$
$$-\chi(1,2)(i\alpha\cdot\nabla_2+\beta m_2)+V(r)[\chi(1,2)-\alpha_\perp\cdot\chi\,\alpha_\perp]$$
$$-V_S(r)\beta\chi\beta\,. \qquad (1)$$

Here we have a 4x4 matrix representation for the 16-components wave function $\chi$, which transforms like

$$\chi_{\alpha\beta}(1,2)\sim\psi_\alpha(1)\psi_\beta^{+}(2),$$

$\alpha_\perp$ is the component of Dirac $\alpha$ perpendicular to $\vec{r}=\vec{x}_1-\vec{x}_2$. Thus, the last two terms on the right hand side of (1) represent the vector and the scalar potentials, respectively. By reducing eq.(1) to the one for a large-large component, as is done in the case of a positronium, one readily finds that the scalar potential gives a spin-orbit term from the Thomas precession which is opposite in sign to the one obtained from the vector-type. Thus a phenomenologically favorable choice will be to take $V(r)=-\alpha_S/r$ and $V_S(r)=Kr$. Then the long range spin-orbit terms tend to reduce the one from the short range vector-type. Furthermore, the long range potential does not contribute to the hyperfine and the tensor interactions. The problem in this scheme is that there is no apparent mechanism in QCD to give a long range term like $\beta\chi\beta$, other than an impossible scalar exchange. There have been attempts[2)] to explain the spin-dependent forces in heavy quarkonia on the basis of QCD. Our approach is slightly different and we will examine directly existence of $\beta\chi\beta$ term. In fact we will show that one can derive a rather definitive wave equation in the confinement range once two basic approximations are made.

These are the large N (N denoting the color dimension) approximation and the linear-flux approximation, which will be defined precisely later. We will find that we obtain two types of potentials which may be called "vector" and "scalar" in a sense different from the above and that there can be no scalar potential of the form $\beta \chi \beta$. Yet the wave equation could yield an appropriate fine structure of the heavy quarkonia.

Our formalism[3] is based on a gauge independent q-$\bar{q}$ operator defied in terms of the quark spinor field q(x) and the gluon field $\vec{A}^a(x)$ ($A_0=0$) like

$$q_{\alpha\beta}(1,2)=N^{-1}Tr^c[U(2,1)q_\alpha(1)q_\beta^+(2)], \tag{2}$$

where $Tr^c$ is the color trace.
U(1,2) is given by

$$U(2,1)=Pe^{ig\int_1^2 \underset{\sim}{\vec{A}}(x)\cdot d\vec{x}}, \qquad (\underset{\sim}{A}\equiv \sum_a A^a \lambda^a/2)$$

where P denotes the path-ordering and the path is taken to be a straight line. The equation of motion for q(1,2) ($t_1=t_2$) was derived in reference 3 and is

$$[q(1,2),H]=(-i\alpha\cdot\nabla_1+\beta m)q(1,2)-q(1,2)(i\alpha\cdot\nabla_2+\beta m)$$
$$+g\int_1^2 d\vec{x}\cdot[\underset{\sim}{\vec{E}}(x)]_{1,2}+\text{magnetic terms}. \tag{3}$$

Here $[\underset{\sim}{E}(x)]_{1,2}$ is an abbreviation for

$$[\underset{\sim}{E}(x)]_{1,2}=N^{-1}Tr^c[U(2,x)\underset{\sim}{E}(x)U(x,1)]q(1)q^+(2)] \tag{4}$$

To obtain a wave equation for a c-number wave function

$$\chi(1,2)=\langle 0|q(1,2)|\rangle,$$

where $|\rangle$ is a hadronic state at rest, we have somehow to make a reduction of the matrix element of $[E(x)]_{1,2}$ and $[B(x)]_{1,2}$. The linear flux approximation mentioned before is

$$\langle 0|[E_\perp(x)]_{1,2}|\rangle=0, \tag{5}$$

stating that components of the electric flux perpendicular to the z-axis, taken to be the line connecting point 1 to 2, vanish. The assumption (5) allows us to write

$$\langle 0|[E_z(x)]_{1,2}|\rangle=\frac{1}{2}\int_{-\infty}^{\infty}\varepsilon(z-z')\vec{\nabla}'\cdot\langle 0|[\vec{E}(x')]_{1,2}|\rangle dz', \tag{6}$$

where $\vec{x}=(x_\perp,z)$ and $\vec{x}'=(x_\perp,z')$. In deriving (6) we assumed

$$\langle 0|[E(x_\perp,\pm\infty)]_{1,2}|\rangle=0 \tag{7}$$

meaning that an amplitude to find an infinitely long, folded electric flux in a hadron should vanish. This would certainly be right as long as $x_\perp=0$. If $x_\perp=0$, as in the case of 1+1 dimension[4)], the folded parts overlap and could cancel themselves, giving a finite value to $[E(\pm\infty)]_{1,2}$. We will always consider the limit $x_\perp\to 0$, so that (7) remains valid. In ref.4, we showed how to apply the gauss' law imposed on physical states,

$$[D_A\cdot E-\rho]|\text{phys}\rangle=0;\ \rho^a=q^+(x)(\lambda_a/2)q(x) \tag{8}$$

to convert equation (6) to

$$\langle 0|\underset{\sim}{E}_z(x)]_{1,2}|\rangle=\frac{1}{2}\int_{-\infty}^{\infty} dz'\,\varepsilon\,(z-z')\langle 0|[\rho\,(x')]_{1,2}|\rangle + \ldots\ldots \qquad (9)$$

The dotted part is produced in pulling the external derivative inside and involves B field in addition to E, thus vanishing in the 1+1 dimension. We can also establish readily that it vanishes in 4-dimension also upon integration of (9) on x from 1 to 2. Noting that

$$\int_1^2 \varepsilon\,(z-z')dz=|\vec{x}_2-\vec{x}'|-|\vec{x}'-\vec{x}_1|, \qquad (10)$$

we find

$$\int_1^2 d\vec{x}\cdot\langle 0|[\underset{\sim}{\vec{E}}(x)]_{1,2}|\rangle= \int_{-\infty}^{\infty} dz'(|\vec{x}_2-\vec{x}'|-|\vec{x}_1-\vec{x}'|)\langle 0|[\rho\,(x)]_{1,2}|\rangle \qquad (11)$$

Using a Fierz transformation[3),4)]

$$[\rho\,(x)]_{1,2}=\frac{N}{2}q(1,x)q(x,2)-\frac{1}{4}[\rho_{\,S}(x),q(1,2)]_+ \qquad (12)$$

where $\rho_{\,S}(x)=\frac{1}{N}q^+(x)q(x)$, we obtain

$$\int_1^2 \langle 0|[\underset{\sim}{\vec{E}}(x)]_{1,2}|\rangle\cdot d\vec{x} =\frac{g\,N}{4}\int_{-\infty}^{\infty} dz(|\vec{x}_2-\vec{x}|-|\vec{x}_1-\vec{x}|)\langle 0|q(1,x)q(x,2)|\rangle. \qquad (13)$$

Again in the sprit of the large N approximation we may drop all hadronic intermediate states except the vacuum in taking the matrix element of (16). Defining the vacuum function

$$S(x_2-x_1)=\langle 0|q(1,2)|0\rangle$$

we obtain finally from (3) a wave equation

$$E\chi(\vec{r})=[-i\alpha\cdot\nabla+\beta m,\chi(\vec{r})]$$
$$+\frac{g^2N}{4}\int_{-\infty}^{\infty}dz[|\vec{r}-\vec{x}|-|\vec{x}|)[S(\vec{x}),\chi(\vec{r}-\vec{x})] \qquad (14)$$

In the above we have taken the system at rest. Also, for brevity we omitted the contribution from the magnetic term, which will be discussed elsewhere.

The vacuum function $S(\vec{r})$ can be expanded in the Dirac matrices like

$$S(r)=S_0(r)-i\alpha\cdot\hat{r}S_1-i\beta\;\alpha\cdot\hat{r}S_2(r)+\beta\,S_3(r), \qquad (15)$$

$S(\vec{r})$ satisfies from (16) an equation corresponding to (17)

$$0=[-i\alpha\cdot\nabla+\beta m,S(\vec{r})]+\frac{g^2N}{8}\int dz[|\vec{r}-\vec{x}|-|\vec{x}|][S(\vec{x}),S(\vec{r}-\vec{x})]. \qquad (16)$$

Writing out the equation in terms of $S_i(r)$ we easily see that we may choose $S_2=0$, which leaves a single equation involving $S_1$ and $S_3$. There is another equation for S which follows from the equation (7), so that in principle we could solve for $S_1$ and $S_3$. Although we have not succeeded in doing that, we could still derive some conclusions from the equation (14) in the following way. We will first consider the case of vanishing quark mass m=0, although this is outside our initial interest. The invariance under the chiral transformation

$$S(\vec{r})\rightarrow S(\vec{r})+i\theta\,[\gamma_5,S(\vec{r})] \qquad (17)$$

would mean

$$S_3(r)=0$$

If on the other hand the symmetry is broken in the vacuum, we will have $S_3 \neq 0$ and the vacuum condensate is given by

$$\langle \bar{q}(0)q(0)\rangle_0 = -4NS_3(0) \tag{18}$$

In the latter case we can easily show from the axial vector divergence relation that the corresponding Ward-Takahashi identity gives

$$\chi_\pi(r) = \frac{i}{\sqrt{2}f_\pi}[\gamma_5, S(\vec{r})] \propto S_3(r). \tag{19}$$

Here $\chi_\pi(r)$ represents the wave function of the massless Goldstone boson. Thus $S_3(r)$ is the pion wave function and will have a finite range consistent with the assumption $S_3 \neq 0$. We may for instance take

$$S_3(r) = S_3(0)e^{-D_\pi r}, \tag{20}$$

although the following argument does not depend on this specific form. Introducing (23) into (17), $D_\pi S_3(|x|)/2S_3(0)$ may be treated like a delta function in the region $r \gtrsim 1/D_\pi$, and we obtain from the $S_3$term a "scalar" potential

$$\frac{1}{2}K_s r[\beta, \chi(\vec{r})] \tag{21}$$

with

$$K_s = \frac{g^2 N}{D_\pi} S_3(0) = -\frac{g^2}{4D_\pi}\langle \bar{q}(0)q(0)\rangle_0 . \tag{22}$$

We called this a scalar potential because it adds to the quark mass term. The $S_1$ term on the other hand contributes a "vector" type as seen in the following way. Assuming that $S_1(r)$ has a form similar to $S_3(r)$, (20), we can treat it like we did $S_3(r)$ and obtain

$$[-i\alpha\cdot\nabla, br\chi(r)] \tag{23}$$

with

$$b=-g^2NS_1(0)/2D_\pi{}^2 \tag{24}$$

Combining it with the Dirac kinetic energy term in (14) and introducing a new wave function

$$\phi=(1+br)\chi \tag{25}$$

we obtain

$$E\phi=[-i\alpha\cdot\nabla, \phi]+V(r)\phi+\frac{1}{2}V_S(r)[\beta, \phi], \tag{26}$$

with

$$V(r)=K_r r/(1+br). \qquad (K_V=EB) \tag{27}$$

and

$$V_S(r)=K_S r/(1+br). \tag{28}$$

The V(r) term, originating from $S_1$, may be called "vector" as it adds to E. We find no scalar-exchange type, $\beta\phi\beta$ term as in (1). The difference between $\beta\phi\beta$ and $[\beta, \phi_2]$ is important because one get no Thomas precession term from the latter.

In the case $m=0$, we proceed in a similar fashion. Let's assume again that $S_1$ and $S_3$ have a finite range D, so that in

$r \gtrsim D^{-1}$ we obtain (21), (22), (23) and (24) with $D_\pi$ replaced by D. Now the $S_1$ and $S_3$ terms must be regrouped into a sum of a Dirac kinetic energy term $[-i\alpha\cdot\nabla+\beta m,\chi]$ and a scalar term $[\beta,\chi]$. We obtain

$$E\phi=[-i\alpha\cdot\nabla+\beta m,\phi]+V\phi+\frac{1}{2}V_s(r)[\beta,\phi] \tag{29}$$

where V and $V_s$ are again given by (27) and (28) with

$$b=-g^2NS_1(0)/2D^2$$
$$K_s=(g^2N/D)[S_3(0)+\frac{m}{D}S_1(0)]. \tag{30}$$

Note that for the perturbative vacuum, for which $S(\vec{r})$ is equal to the positive energy projection operator, we have D=m, and $S_1(0)/S_3(0)\sim -1$, hence $K_s=0$. In order to have a dominance of the scalar potential in the confinement range, we must have

$$S_3/S_1 \gg 1 \tag{31}$$

It would seem that the vacuum must be significantly different from the perturbative state in order to have the relation (31). Once this condition is satisfied, we will have $V=-\alpha_s/r$, and $V(s)=K_s r$, so that there is no contribution of a spin-orbit term from the long range potential. Contrary to the conventional scheme, the scalar part has also the magnetic interaction, which were omitted here. The over all tendency of the spectrum that our equation (32) gives seems to be good.

The author would like to thank members of the theoretical high-energy groups at University of Helsinki, University of Tokyo and Nihon University for their hospitality in the fall of 1985 when this work was done. I thank Hitoshi Ito and

Kenji Yamada for useful comments.

References

1)A.B.Henriques, B.H.Kellett and R.G.Moorhous,
Phys.Lett.64B,85(1976)
D.Gromes, Nucl.Phys.B131,B131(1977)
L.H.Chan, Phys.Lett.71B,422(1977)
H.J.Schnitzer,Phys.Lett.76B,461(1978);Phys.Rev.D18,3482(1978)

2)E.Eichten and F.L.Feinberg,
Phys.Rev.Lett.43,1205(1979);Phys.Rev.D23,2724(1981)
D.Gromes, Z.f.Physik,C26,401(1984)
W.Buchmuller,Phys.Lett.112B,479(1982)

3)H.Suura, Phys.Rev.D20,1412(1979)

4)H.Suura, University of Minnesota preprint (1985)

# COLOR-CONFINEMENT, CHIRAL SYMMETRY BREAKING AND THE STRUCTURE OF HADRONS

Tomoya Akiba
Department of Physics
Tohoku University, 980 Sendai, Japan

The quark model picture of hadrons (especially, light-quark hadrons not including heavy quarks) has changed from time to time and is continuously developing, although it is believed that the ultimate description will be provided by quantum chromodynamics (QCD). In the early 1960's the non-relativistic quark model was used to calculate hadronic properties and achieved considerable success.[1] Although a better description[2] of the model was given after the introduction of QCD, the question remained why the quarks inside a hadron could be treated as non-relativistic particles. Subsequently the MIT bag model[3] and its variants, the chiral bag models,[4] were introduced. There quarks were regarded as almost massless Dirac particles confined in a bag. This description appeared to fit well to the observations in deep inelastic scattering experiments. Neverthless, hadronic properties calculated from these models were at best comparable to the achievements of the non-relativistic model. In the last a few years we witnessed the prevalence of the Skirmion model[5] that regards baryons as solitins.

It is evident that each model emphasizes only one side of the real hadrons and has its own limit of application. Perphaps, their limits can be characterized by different scales. It is extremely important to reveal existence of such different scales in QCD.

The two essential ingredients in QCD are color-confinement and spontaneous breakdown of chiral symmetry ($\chi$SB). It seems to me that the most important as well as urgent problem is to understand possible interplay between them. Only after that some significant progress can be expected in understanding the hadron structure.

If one considers the confinement alone, there has been already a remarkable progress from the lattice gauge theory. That is, Monte Carlo simulations[6] yielded clear evidences for existence of the color-flux squeezed like a string between $q$ and $\bar{q}$ at rest. Also, by the effective action calculated in the one-loop approximation one can show that the QCD vacuum is likely to have the property to support bags which confine the color-flux together with $q\bar{q}$.[7]

On the other hand lattice gauge theories have the well-known difficulty in investigating $\chi$SB. Because the doubling problem[8] prevents us from putting chiral fermions on lattices straghtforwardly. Fortunately, the qualitative behavior of $\chi$SB can be understood by the Schwinger-Dyson equation. Previously most of works solved the SD equation incorporating only the one-gluon exchange potential (OGEP). To be more quantitative, one must include the confining potential in addition to OGEP. Use of the Coulomb gauge[9] is likely to make such improvement feasible and some progress can be expected in the approach with the SD equation.

The hadronic structure looks rather different, depending on the scales with which we prove hadrons. The non-relativistic quark model may be the picture against the probe of a bit smaller scale than the hadron size. The quarks in this scale are properly called "constituent quarks". They are different from "current quarks" which emerge under probes of finer scale. My analysis[10] suggested the picture of a constituent quark as a current quark confined in a small bag of radius 0.6−0.7 fm. To my regret, this picture does not yet given firm links with QCD.

Finally let us turn our attention to the internal structure of pion (in general, octet ps mesons). The pion is simultaneously a composite of $q\bar{q}$ and the Nambu-Goldstone boson, the latter being regarded as fluctuations of the $q\bar{q}$ condensate vacuum. Because of the peculiarity one expects different structure of pion compared to other hadrons. At least two mass scales $\Lambda_c$ and $\Lambda_{\chi SB}$ are pertinent to the pion. Here $\Lambda_c$ represents the characteristic mass scale for the confinement, while $\Lambda_{\chi SB}$ the one for $\chi$SB. Some phenomenological analysis[11] suggested $\Lambda_c/\Lambda_{\chi SB} \lesssim \frac{1}{3}$. Is this consistent with the fact that the SU(3) lattice gauge theory at finite temperature predicted almost simultaneous occurrence of deconfinement and $\chi$SB-restoration?[8]

We may study the chiral condensate with the SD equation, incorpolating some confining potential and the perturbative OGEP. Through the Ward-Takahashi identity the condensate function is related to soft-pion's wave function. If the OGEP is such that it alone cannot support the $q\bar{q}$ condensate, one can easily understood the simultaneous occurrence of deconfinenent and $\chi$SB-restoration. However, the question is: What does $\Lambda_{\chi SB}$ mean?

Our interpretation is that $\Lambda_{\chi SB}^{-1}$ represents an effective pion size. If the OGEP gives appreciable contribution to the $q\bar{q}$ condensate in the range of $q\bar{q}$ relative momentun up to, say, 1 GeV/c, then the effective size may be regarded as small compared to other hadrons, which in turn results in a small value of $\Lambda_{\chi SB}^{-1}$. The mean squared radius of pion's charge distribution is estimated[12] to be $\langle r^2\rangle_\pi = 0.46\ \mathrm{fm}^2$. The smallness of this quantity is not at variance with our speculation, though it is not sensitive to the small-distance behavior of the wave function. Perphaps, the detailed measurements of the x-distribution of the valence quarks in the pion structure function may yield valuable information. Also, the mass difference $\mathrm{m}_{\pi^\pm}^2 - \mathrm{m}_{\pi^0}^2$ is sensitive to the internal structure of pion.

We hope that in the near future our qualitative argument will turn into a quantitative one.

My interest in the bag model was stimulated by the lecture of Professor Gyo Takeda, primarily addressed to graduate students and experimental physicists, in the fall of 1980. I am grateful to him for his valuable suggestions throughout my research. It is a pleasure to express my thanks to Professor Gyo Takeda at the celebration of his sixtieth birthday.

## References

1) F. E. Close, *An Introduction to Quarks and Partons* (Academic Press, New York, 1982).
2) A. De Rújula, H. Georgi and S. L. Glashow, Phys. Rev. D12 (1975) 147.
3) A. Chodos, R. L. Jaffe, K. Johnson, C. B. Thorn and V. F. Weisskopf, Phys. Rev. D9 (1974) 3471.
4) T. Inoue and T. Maskawa, Prog. Theor. Phys. 54 (1975) 1833.
A. Chodos and C. B. Thorn, Phys. Rev. D12 (1975) 2733.
5) T. H. R. Skyrme, Proc. Roy. Soc. A260 (1975) 127.
6) M. Creutz, Phys. Rev. Lett. 45 (1980) 313.
7) S. L. Adler and T. Piran, Phys. Lett. 113B (1982) 405; ibid. 117B (1982) 91.

T. Akiba and F. Takagi, Phys. Lett. 131B (1983) 187.

8) J. B. Kogut, Rev. Mod. Phys. 55 (1983) 775.

9) J. R. Finger and J. E. Mandula, Nucl. Phys. B199 (1982) 168.

T. Akiba, Prog. Theor. Phys. (L) 74 (1985) 641.

10) T. Akiba, Phys. Lett. 109B (1982) 477; Prog. Theor. Phys. 67 (1982) 1822.

11) A. Manohar and H. Georgi, Nucl. Phys. B234 (1984) 189.

12) S. Dubnicka, V. A. Meshcheryakov and J. Milko, J. Phys. G7 (1981) 605.

# Vector Mesons Forever!

Takeshi Ebata
College of General Education
Tohoku University, 980 Sendai, Japan

My acquaintance with Professor Gyo Takeda began around 1960. At that time he was a professor at the Institute for Nuclear Study, University of Tokyo, and I was a graduate student at Tohoku University. In these days, the very existence of the rho-meson was a topic in the seminars, and the experimental high energy physics in Japan was at its infancy with the first accelerator, INS electron synchrotron. Itabashi and myself calculated the cross-section for the process $\gamma p \to \pi^0 \pi^0 p$ as one possible experiment with the INS machine, where we have used the notion that the J=1 hadronic (two pion) state may be related to the electromagnetic current. Later we have expanded the idea to the calculation of the process $\pi N \to \pi \Delta$ and the subject "isobar production in boson-nucleon collision" formed a main part of my thesis[1]. I recall that my interest in the rho-meson was stimulated by the seminar given by Professor Takeda who frequently visited our university. In the mean time, Stodolsky and Sakurai explained the reaction $KN \to K\Delta$ with the vector dominance theory advocated by Sakurai[2]. And the success stimulated the development of the vector dominance theory. Though our approach lacked the philosophy, which is quite important, the method was identical with theirs. The reason I recall the old story now is to remind the fact that Professor Takeda has long been interested in the vector mesons and his students benefited from his interest.

Later, with the development of the S-matrix theory, I was interested in the kinematical structure of the helicity amplitudes involving these neutral vector mesons such as $\rho$ and $\omega$, and its relation with the corresponding photo-processes. Consider the process

$$\rho^0(m) + K^+ \rightarrow \rho^0(m) + K^+ \tag{1}$$

as an example. In the above, the squared four-momentum of the vector meson is denoted as $m^2$, wheras the observed mass is $m_V^2$. It is shown that the helicity amplitudes containing n longitudinal helicity states can be demanded to vanish as $m^n$ when the external $\rho$-meson "mass" m is varied to zero[3]. Together with the analyticity requirement, the aforementioned condition for the longitudinal helicity amplitudes reduces to the gauge invariance requirement for the Compton process

$$\gamma + K^+ \rightarrow \gamma + K^+. \tag{2}$$

This property guaratees the validity of the vector dominance model in the helicity formalism. Since the vector dominance model is known to be quite successful in the "soft region", the situation described above may not be surprising. It is amazing, however, to note that there exists a massive particle, whose helicity amplitudes can be smoothly connected to the zero mass particle (photon) amplitudes, since the little groups for massive and massless particles are completely different.

The discussion given above may be expressed in a different way. A photon which propagates between the hadronic and the leptonic currents has to be represented by

$$1/q^2 - 1/(q^2 - m_V{}^2)$$
$$= -m_V{}^2/(q^2 - m_V{}^2), \qquad (3)$$

instead of the propagator $1/q^2$. (We ignore the contributions from other $\rho$'s for simplicity.) The relative weight of the $\rho$-pole to the photon pole is definitely prescribed in the vector dominance model. That the reaction (1) can be related to reaction (2) in helicity formalism is nothing but the kinematical manifestation of the dynamical relation (3).

We notice, however, that the possibility of continuing the external "mass" m to zero in reaction (1) is intimately related to the crossing and analytic properties of the amplitudes. Consider the charge exchange reaction

$$\rho^+(m) + K^0 \to \rho^0(m) + K^+. \qquad (4)$$

In this case the continuation of the mass "m" to zero in the helicity formalism is not possible because of the wrong crossing property.[3)] Dynamically, this is because of the exchanged $\rho$-pole due to the $\rho^+\rho^-\rho^0$-vertex. (Note that the exchanged $\rho$ has a pole at $t=m_V{}^2$, which is kept fixed.)

Thus, the possibility of the continuation of the "mass" m to zero is achieved at the expense of the isospin symmetry breaking. This point may be seen more easily from the following consideration. The vector dominance relation implies that the neutral rho is always pure spin one object, that is

$$\partial^\mu \rho_\mu{}^0(m) = 0 \qquad (5)$$

for arbitrary m, whereas for the charged rho, the corresponding relation can only be

$$(\partial^{\mu} - ie\, A^{\mu})\rho^{+}_{\mu}(m) = 0. \qquad (6)$$

On one hand, the simple recollection of the humble personal activity is my tribute to Professor Takeda for his continual interest and encouragement. On the other hand, it seems to me that the problem of the vector mesons has fresh implication today. These vector bosons act almost like gauge bosons, yet, according to the present day belief, they are not connected with any gauge principle, which exists a priori. They are massive because those quarks which compose them, are massive. The approximate conservation of the isospin current, for example, is guaranteed by the existence of the photon through the vector dominance relation; the existence of the photon, however, inevitably breaks the isospin symmetry. Though the standard Weinberg-Salam theory is perfectly in agreement with the experiment (to the extent of its accuracy), we are in a tantalizing situation[4]. Are the intermediate bosons W and Z real gauge bosons or mere analogues of the $\rho$-meson? Does the Higgs boson exist and has mass of the order of $10\sim100$ GeV? And, so on. On these respects, recent suggestion[5] that the $\rho$-meson might be "a dynamical gauge boson of a hidden local symmetry", is quite interesting. We wish that further clarification would give more insight toward the understanding on the structure of the Nature. Why the Nature provides us so many approximate symmetries with few exact symmetries?

In conclusion, I hope that Professor Takeda will give us useful suggestions in the coming years as he did.

References

1). K.Itabashi and T. Ebata, Prog. Theor. Phys. 28(1962) 915;

K. Itabashi and T. Ebata, Prog. Theor. Phys. 30(1963) 505;

T. Ebata, Prog. Thor. Phys. 31(1964)197.

2). J. J. Sakurai, Ann. of Phys.11(1960)1;

L. Stodolsky and J. J. Sakurai, Phys. Rev. Letters 11(1963)90.

3). T. Ebata, T. Akiba and K. E. Lassila, Prog. Thor. Phys.44(1970)1684;

T. Akiba, M.Sakuraoka and T. Ebata, Nucl. Phys. B31 (1971)381.

See also, F. Arbab and R. C. Brower, Phys. Rev.181 (1969)2124.

4). R. Schwitters in Proceedings of the 1985 International Symposium on Lepton and Photon Interactions at High Energies (Kyoto).

5). M. Bando, T. Kugo, S. Uehara, K. Yamawaki and T. Yanagida, Phys. Rev. Letters 54(1985)1215.

# MULTIPLE PRODUCTION IN HADRON-NUCLEUS COLLISIONS AT HIGH ENERGIES

Fujio Takagi
Department of Physics
Tohoku University, 980 Sendai, Japan

## I. INTRODUCTION

Physics of multiple production in hadron(h)-hadron(h) collisions reached its peak in early '70s soon after the proposal of the scaling[1] or the limiting fragmentation[2] and the Regge-Mueller formalism[3] of the inclusive reactions. It then declined in inverse proportion to the establishment of the so-called standard theories based on the nonabelian gauge fields. However, a small revival took place in mid '70s when high precision accelerator data became available for multiple production in hadron(h)-nucleus(A) collisions at high energies and some interesting models were proposed thereby.[4),5)] Multihadron physics enthusiasts have then made a lot of efforts to develop this field of physics.[6] More than a decade has passed, and the present status of h-A physics is rather unsettled. This is because the main interests have turned to the new aspects of multiple production in h-h collisions at $\sqrt{s}$=540 GeV (CERN SPS Collider) and beyond or to the physics of quark-gluon plasma which is expected to be formed in nucleus(A)-nucleus(A) collisions at sufficiently high energies. It seems, therefore, timely to give a short (mainly theoretical) review of high energy h-A physics in an attempt of summarizing what have been established and what will be expected in near future, in connection with the future prospects of h-h and A-A physics.

Only a limited number of references will be cited as this is not a comprehensive review.

## II. FACTS AND CONTROVERSIES

### 1. Formation Length

It has been claimed repeatedly that multiple production in h-A collisions is useful to study the space-time structure of production processes. This is simply because one can control the target thickness, i.e., the longitudinal size of the region of the initial interaction by varying the target mass number A. Though the way of extracting information on the space-time structure from the A-dependence of appropriate observables is highly model-dependent, there is a consensus on the statement that most of hadrons are produced outside nuclei in high energy h-A collisions. This is an important consequence obtained from the so-called formation zone concept. It is easily seen in various ways that a finite space-time region (the formation zone) is necessary for production of a hadron in its rest frame. A typical distance, i.e., the formation length needed for production of a hadron in its rest frame is $\ell_0 \simeq 1$ fm on the average. Here, $\ell_0$ reflects simply the size of the hadron. The typical proper formation time is hence $\tau_0 \simeq 1$ fm/c. The formation length in the frame where the produced hadron has a large momentum becomes larger than a typical nuclear radius because of the Lorentz time dilation.

The large formation length implies the absence of large scale intranuclear cascade by the secondary particles. The absence of intranuclear cascade then implies that the A-dependence of the mean multiplicity in h-A collisions is weak in agreement with the experimental fact that

$$\langle n\rangle_{hA} / \langle n\rangle_{hN} \simeq A^{0.1\sim0.2} , \qquad (2.1)$$

where the subscript N refers to a nucleon.

The idea was realized in many different models, e.g., the collective tube model,[5] the additive quark model,[7] the dual parton model,[8] the multichain model[9] and so on. Such a diversity is mainly due to lack of a unique understanding of multiple production in h-h collisions. We do not enter into detailed comparison of various models.

## 2. Nuclear Transparency and Nuclear Attenuation

Nuclear transparency and nuclear attenuation represent both sides of the same phenomenon, i.e., the weak A-dependence of the leading particle effect. The leading particle effect is characterized by the inelasticity $\eta$ which is the fraction of the beam energy used for inelastic interactions:

$$\eta = 1 - E_{\text{leading particle}} / E_{\text{beam}} . \qquad (2.2)$$

The $\eta$ is about 0.5 in inelastic proton-proton collisions, and for proton-nucleus collisions, it has been known to be almost independent of or only weakly dependent on A. This is called the nuclear transparency and is intimately related to the large formation length discussed in the preceding subsection.

High precision experiments[10] have shown, however, that the spectra of fast forward hadrons decrease considerably as A increases. This nuclear attenuation of leading particles implies that the inelasticity $\eta_{hA}$ is an increasing function of A. At present, theoretical realization of this effect is highly model-dependent. For example, in the multichain model[9], the leading cluster loses its energy-momentum gradually during the course of sequential multiple collisions, while the additive (or constituent) quark model in its extreme version supposes that the wounded quarks almost stop in the central rapidity region while the spectator i.e., the leading quarks preserve their initial momenta.[11]

3. Abnormal Nuclear Enhancement at High Transverse Momenta

Cross sections for inclusive reactions $h+A \to c+$ anything in a high transverse momentum region exhibit a striking nuclear effect.[12] If one parametrizes the A-dependence in the form

$$E \frac{d^3\sigma_{hA\to cX}}{d^3\vec{p}} \propto A^{\alpha(p_T)} , \tag{2.3}$$

the exponents $\alpha(p_T)$ for $h=\pi^{\pm}, p$ and $c=\pi^{\pm}$, $K^{\pm}$, $p$, $\bar{p}$ exceed a naive expectation $\alpha(p_T)=1$ and reaches around $1.1 \sim 1.3$. This abnormal nuclear enhancement is strongest for $c=K^-$, $p$ and $\bar{p}$ when $h=p$. Though the effect is generally believed to be consistent with the multiple hard scattering picture, the dependence on the particle species has not been well elucidated yet.

High $p_T$ jet processes $h+A\to$ jet + anything show a much stronger effect.[13] We suggested that contamination of multiple jet production due to simultaneous hard collisions may be important.[14] Unfortunately, no further experiment has been done to disentangle this problem.

## 4. Implication from Lepton Pair Production

The A-dependence of the reaction $h+A \rightarrow \mu^{+}+\mu^{-}+\text{anything}$ is consistent with $A^1$ behavior for a wide range of the phase space of the μ pairs.[15] One might consider this result rather trivial and uninteresting because a nucleus is very transparent for the electromagnetic interaction and hence the cross section should be proportional to the number of the constituents. However, it actually implies an important fact that the momentum distribution of incident hadrons or their constituents does not change significantly when they traverse the target nuclei. Change of the momentum distribution and the production of secondary partons (either quarks, antiquarks or gluons) take place mainly outside the target nuclei after the passage of incident hadrons. This conclusion is perfectly consistent with the formation length argument given in subsection 1.

## 5. Cumulative Production

In a collision of an incident hadron on another nucleon at rest, a nucleon cannot be emitted in the backward direction because of energy-momentum conservation. Such a "kinematically forbidden" reaction takes place in h-A collisions.[16] Many nucleons in a target nucleus must be involved in such a process. Effects from those nucleons must be accumulated in such a cooperative way as to produce a backward hadron with a significant momentum. The phenomenon has been thereby called the cumulative production.[16)]

Possible theoretical explanations can be classified into two categories. The major opinion is that the effect is due to a short range correlation such as a high momentum tail of the nucleon Fermi motion, a direct many nucleon correlation of short range nature or multiquark states already present in stationary nuclei.[17)] We would like to reserve a minor opinion that the effect may be mainly due to a long range correlation of nonperturbative nature such as quark recombination[18)] or color excitation.[19)]

## 6. Collective vs. Noncollective Models

There are two major classes of models for multiple production in h-A collisions. One is the collective models and another is the noncollective ones. In a noncollective model, multiple production in h-A collisions at some incident energy is described by an incoherent sum of "elementary" collisions at the same or the lower energies. The distribution of hadrons in the elementary collision may be estimated from h-N data. Therefore, h-A collisions at some energy are predicted to be similar to (a sum of) h-N collisions at the same or the lower energies. The situation is opposite in the collective models where many nucleons or many quarks (and gluons) from many intranuclear nucleons are supposed to react collectively against the incident hadron. In this case, the effective mass of the target becomes larger than that in h-N collisions, thus raising the effective energy in h-A collisions in comparison with the same in h-N collisions at the same incident energy. In this sense, multiple production in h-A collisions is expected to be similar to that in h-N collisions at higher energies. The difference between two classes of models is conceptually clear. Actual distinction by experiment is, however, not so easy. We proposed that the A-dependence of the particle ratios such as $K^-/\pi_{ch}$ or $\bar{p}/\pi_{ch}$ in the central rapidity region of p-A collisions is one of the best quantities to distinguish between the two classes of models.[20] Collective models predict that those ratios increase as A increases, while the noncollective models predict the opposite.[20] Data sufficiently good for this purpose is not available yet.

## III. FUTURE PROSPECTS

Though the subjects reviewed in Sec.II are rather limited, they already prove that high energy h-A physics has brought in many important information which cannot be obtained from h-h physics. Knowledge and experience of h-A collisions are in fact very useful to calculate such important quantities as the nuclear stopping power or the energy density to be achieved in A-A collisions.[21] The main purpose of such calculations is to assess the possibility of formation of a new state of matter called quark-gluon plasma.

The main obstacles to the further development of high energy h-A physics are, to my opinion, (i) lack of sufficiently many and sufficiently accurate experimental data, (ii) lack of sufficient understanding of h-h physics, and (iii) lack of sufficient information on low energy nuclear physics such as the amount of the high momentum tail of the nucleon Fermi motion. In addition to conventional experiments, special experiments that aim at resolution of main theoretical controversies are highly desirable. The search for the collective interaction in h-A collisions is one of the most important subjects. If it is found, it may be a precursor of quark-gluon plasma or even quark-gluon plasma itself. Poor understanding of multiple production in h-h collisions is not so serious. This is indeed one of the reasons why one should study h-A collisions. New information from h-A physics will be useful to disentangle open questions in h-h physics and in fact it has been so. Poor understanding of the Fermi momentum distribution is more serious. It has hampered clear theoretical understanding of such important phenomena as the cumulative production[16)] and the abnormal A-dependence of the nuclear structure functions called the EMC effect.[22)] Therefore, the determination of the Fermi momentum distribution, in particular, in the high momentum region has become one of the most important problems of "low energy ? " nuclear physics for further development of high energy h-A (and A-A) physics.

In high energy h-h physics, quantum chromodynamics has been most useful and successful in high $p_T$ regions. On the contrary, low $p_T$ h-h physics is still in very phenomenological stage reflecting the complexities and difficulties of nonperturbative aspects of quantum chromodynamics such as confinement and chiral symmetry breaking.[6] Multiple production of hadrons, i.e., production and subsequent hadronization of a system with many quarks, antiquarks and gluons should certainly be dominated by those nonperturbative dynamics. The word "hadronization" already foresees a kind of phase transition or something like that. Quark-gluon plasma may have already been produced in h-h collisions at CERN SPS Collider or newer Fermilab Tevatron. It is probable that everybody has simply overlooked it. In this sense, the new phenomena such as the breaking of the KNO scaling and the strong correlation between average $p_T$ and multiplicity found in $\bar{p}p$ collisions at SPS Collider are very important.[23] In connection with these and other phenomena, establishing a unified way of describing high and low $p_T$ physics is one of the most urgent problems assigned to theorists.

It is now certain that h-h, h-A and A-A physics (and also physics of multiple production in lepton-lepton, lepton-hadron and lepton-nucleus collisions) are complementary one another and form a big arena of future high energy hadron physics.

<u>References</u>

1) R.P. Feynman, Phys. Rev. Lett. <u>23</u> (1969) 1414.

2) J. Benecke, T.T. Chou, C.N. Yang and E. Yen, Phys. Rev. <u>188</u> (1969) 2159.

3) A.H. Mueller, Phys. Rev. <u>D2</u> (1970) 2963.

4) See, for example, P.M. Fishbane and J.S. Trefil, Phys. Lett. 51B (1974) 139.
K. Gottfried, Phys. Rev. Lett. 32 (1974) 957.

5) G. Berlad, A. Dar and G. Eilam, Phys. Rev. D13 (1976) 161.
S. Fredriksson, Nucl. Phys. B111 (1976) 167.
F. Takagi, Lett. Nuovo Cim. 14 (1975) 559.

6) See, for example, a comprehensive review by K. Fialkowski and W. Kittel, Rep. Progr. Phys. 46 (1982) 1283.

7) A. Bialas, W. Czyz and W. Furmanski, Acta Phys. Pol. B8 (1977) 585.
V.V. Anisovich, Yu.M. Shabelski and V.M. Shekhter, Nucl. Phys. B133 (1978) 477.
N.N. Nikolaev and S. Pokorski, Phys. Lett. 80B (1979) 290.
A. Dar and F. Takagi, Phys. Rev. Lett. 44 (1980) 768.

8) A. Capella and J. Tran Than Van, Z. Phys. C - Particles and Fields 10 (1981) 249.

9) K. Kinoshita, A. Minaka and H. Sumiyoshi, Prog. Thoer. Phys. 63 (1980) 928.

10) See, for example, D.S. Barton et al., Phys. Rev. D27 (1983) 2580.

11) F. Takagi, Z. Phys. C - Particles and Fields 28 (1985) 439.

12) J.W. Cronin et al., Phys. Rev. D11 (1975) 3105.
D. Antreasyan et al., Phys. Rev. D19 (1979) 764.

13) C. Bromberg et al., Phys. Rev. Lett. 42 (1979) 1202; ibid 43 (1979) 1057(E).

14) F. Takagi, Phys. Rev. Lett. 43 (1979) 1296.

15) J. Badier et al., Phys. Lett. 104B (1981) 335.

16) A.M. Baldin et al., Yad. Fiz. 20 (1974) 1201. [Sov. J. Nucl.

Phys. 20 (1975) 629.]

17) See, for example, L.L. Frankfurt and M.I. Strikman, Phys. Rep. 76 (1981) 215.

18) F. Takagi, Phys. Rev. D19 (1979) 2612.

19) B.Z. Kopeliovich and F. Niedermayer, Phys. Lett. 117B (1982) 101.

20) F. Takagi, Prog. Theor. Phys. 66 (1981) 964.

21) W. Busza and A.S. Goldhaber, Phys. Lett. 139B (1984) 235.
H. Sumiyoshi, S. Daté, N. Suzuki, O. Miyamura and T. Ochiai, Z. Phys. C - Particles and Fields 23 (1984) 391.
S. Daté, M. Gyulassy and H. Sumiyoshi, preprint INS-Rep-535, March 1985.

22) J.J. Aubert et al., Phys. Lett. 123B (1983) 275.
A. Bodek et al., Phys. Rev. Lett. 50 (1983) 1431.

23) G.J. Alner et al., (UA5 Collaboration), Phys. Lett. 138B (1984) 304.
G. Arnison et al., (UA1 Collaboration), Phys. Lett. 118B (1982) 167.

# II. NUCLEI AND ATOMS

# EXCHANGE CURRENT CONTRIBUTIONS TO ISOSCALAR MAGNETIC MOMENTS

A.Arima, W.Bentz and S.Ichii

Department of Physics, Univ. of Tokyo
Bunkyo-ku, Hongo 7-3-1, Tokyo 113
JAPAN

[This paper is dedicated to Professor Gyo Takeda for his 60th anniversary.]

## 1. INTRODUCTION

It is nowadays well established that the second order core polarization (tensor correlation) plays a very important role in the explanation of the deviations of isoscalar magnetic moments and the expectation value of the spin from the Schmidt values [1],[2]. Indeed, a glance at our table 2 shows that for nuclei with an LS-closed core plus or minus one nucleon the experimentally observed deviations can be explained quite satisfactoryly on account of core polarization alone. Thus one might expect that the contributions from meson exchange currents should be small, and explicit calculations [1],[2] seem to support this view. However, there have been two quite independent suggestions recently that there might be further appreciable contributions to isoscalar magnetic moments:
The first one comes from the presently fashionable relativistic $\sigma\omega$-model in the mean field approximation [3],[4], where the Dirac part of magnetic moments becomes enhanced roughly by a factor $m/m^*$, where m is the free nucleon mass and $m^*$ is the effective nucleon mass. Due to the smallness of $m^*$ in these models this results in a large enhancement being in conflict with experiment. The second suggestion [5],[6] is based on the more traditional exchange current approach: For certain potential models it has been shown that the exchange current obtained from a gauge transformation of the spin-orbit part leads to a large quenching of the effective isoscalar spin g-factor.

In sect.2 of this work we investigate more closely these two recent suggestions and their connection to exchange current calculations performed so far. By using arguments based on gauge invariance we will see that both methods discussed above seem to omit certain important contributions. In sect.3 we will discuss exchange current contributions to isoscalar magnetic moments more quantitatively. The actual calculations will be carried out within a one-boson exchange model for the exchange current operator. We will see that exchange current processes in some cases may give important contributions, whose actual size, however, turns out to be quite model dependent.

## 2. CONSTRAINTS IMPOSED BY GAUGE INVARIANCE

### 2.1 The angular momentum g-factor $g_1$ and the $\sigma\omega$-model.

Here we consider the nucleus as infinitely extended nuclear matter plus one odd nucleon at the Fermi surface. On account of gauge invariance alone one can show [7] that the electromagnetic current of the odd nucleon with momentum $\vec{p}$ becomes for zero momentum transfer

$$\vec{j}(q=0) = \frac{\vec{p}}{\varepsilon_p}\frac{1+\tau^z}{2} - \tau^z\hat{p}\,\frac{p^2}{3\pi^2}\,(f_1-f_1') \qquad (2.1)$$

Here $\varepsilon_p$ is the quasiparticle energy and $f_1$ and $f_1'$ are two Landau-Migdal parameters defined in ref.[7]. The point relevant for our discussion is that eq.(2.1) gives the following isoscalar angular momentum g-factor

$$g_1^{(0)} = \frac{1}{2}\,\frac{E_p}{\varepsilon_p} \quad . \qquad (2.2)$$

Here $E_p=\sqrt{m^2+\vec{p}^{\,2}}$ is the free nucleon energy. For a free nucleon we have $g_1^{(0)}=0.5$. Since $E_p \gtrsim \varepsilon_p \approx m$ , eq.(2.2) implies that $g_1^{(0)}$ becomes slightly enhanced due to the interactions. This enhancement cannot exceed a few percent due to the fact that the nucleon separation energy $E_p-\varepsilon_p$ is small.

Let us now discuss the importance of this very general result using the currently fashionable relativistic $\sigma\omega$-model in the mean field approximation as an example: In that model the quasiparticle spectrum has the form

$$\varepsilon_p = E_p^* + \Sigma_v \qquad (2.3)$$

with $E_p^* = \sqrt{\vec{p}^2 + m^{*2}}$ with the effective mass $m^* = m + \Sigma_s$ , and $\Sigma_s$ and $\Sigma_v$ are the scalar and vector parts of the nucleon self energy due to $\sigma$-meson and $\omega$-meson exchange, respectively. Calculations of magnetic moments done so far within this framework [3],[4] have approximated the full vertex in the quasiparticle current by the free vertex,i.e;

$$\vec{j}_0(q=0) = \frac{m}{E_p}\, \bar{f}(\vec{p})\vec{\gamma}\, \frac{1+\tau^z}{2} f(\vec{p}) = \frac{\vec{p}}{E_p^*}\, \frac{1+\tau^z}{2} , \qquad (2.4)$$

where $f(\vec{p})$ is the quasiparticle spinor. Note that the $\omega$-meson does not contribute in eq.(2.4). From eq.(2.4) we would obtain a huge enhancement factor $E_p/E_p^*$ for both $g_l^{(0)}$ and $g_l^{(1)}$. A similar enhancement is obtained for the Dirac part of the spin g-factors. It is now the role of the vertex corrections to cut down the enhancement factor $E_p/E_p^*$ for $g_l^{(0)}$ to $E_p/\varepsilon_p$, as required by eq.(2.2). This vertex correction, which together with the mean-field self energy satisfies the Ward identity, is the RPA-type vertex shown in fig.1.

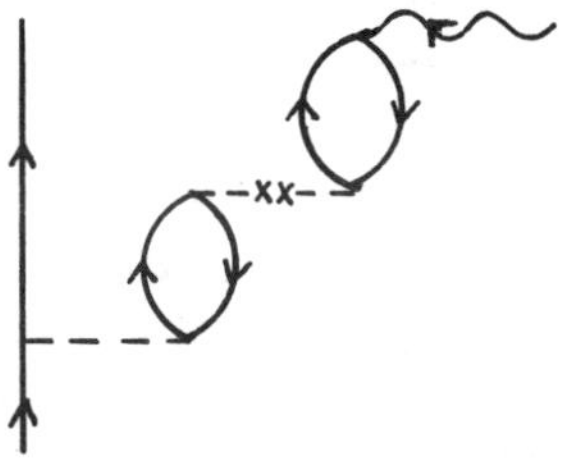


Fig.1 RPA-type vertex correction. The crosses indicate an arbitrary number of bubbles.

For q=0 (note that we first set $\vec{q}=0$ and then $q_0=0$) there survive only nucleon-antinucleon bubbles connected by $\omega$-meson propagators. If we use typical values obtained in the $\sigma\omega$-model, the inclusion of the vertex correction fig.1 changes the enhancement factor $E_p/E_p^* \approx 1.65$ into $E_p/\varepsilon_p \approx 1.08$. It is thus important to note that both in the energy (2.3) and in the current (2.1) we have to have similar cancellations between contributions from σ-meson and ω-meson exchange. In lowest order perturbation theory the corrections to the free nucleon current implied by eq.(2.4) and the vertex correction fig.1 are shown in fig.2a and 2b. Our above arguments imply strong cancellations between these two contributions. In sect.3 we will see that by a usual exchange current calculation one automatically includes both of these two diagrams. This gives an explanation for the contradicting results for isoscalar magnetic moments based on the exchange current approach [1], [2], and the σω-model [3],[4]. The importance of the vertex correction fig.1 has also been pointed out recently by Kurasawa and Suzuki [8].

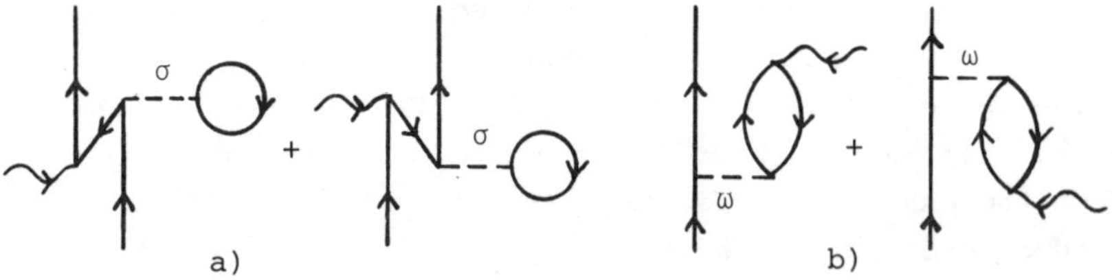


Fig.2 Corrections to the free nucleon current.

## 2.2 Two-body isoscalar exchange currents.

In order to make contact to the second suggestion [5],[6] mentioned in the introduction, let us consider here the requirements of gauge invariance for two-body exchange currents. If one defines a two-body exchange current $J^{\mu}$ via Feynman diagrams in a proper way, it satisfies the following relation [9]

$$q_{\mu}J^{\mu}(q) = [\rho, V] \qquad (2.5)$$

with the commutator defined by

$$[\rho,V] =$$

$$\rho^{(1)}(p_1', p_1'-q)V(p_1'-q, p_2'; p_1, p_2) + \rho^{(2)}(p_2', p_2'-q)V(p_1', p_2'-q; p_1, p_2)$$

$$-V(p_1', p_2'; p_1+q, p_2)\rho^{(1)}(p_1+q, p_1) - V(p_1', p_2'; p_1, p_2+q)\rho^{(2)}(p_2+q, p_2).$$

Here $\rho^{(i)}$ are the one-body charge densities in momentum space and V is the two-body potential. The momenta $p_1, p_2 (p_1', p_2')$ refer to the two nucleons in the initial (final) state. For the calculations in sect.3 we are interested in the isoscalar exchange current in order $1/m^2$ due to one-boson exchange. The potential then takes the form

$$V = V_{loc}[1] + V_{loc}[1/m^2] + V_{mom}[1/m^2] + V_{LS}[1/m^2] \ . \quad (2.6)$$

Here $V_{loc}$ is the local part of the potential, $V_{mom}$ is a momentum-dependent but spin-independent term, and $V_{LS}$ is the spin-orbit potential. In eq.(2.6) we also indicated the relativistic order, being either 1 or $1/m^2$. Similarly, the one-body charge density is written as $\rho = \rho[1] + \rho[1/m^2]$, and eq.(2.5) for the isoscalar part becomes in order $1/m^2$

$$\vec{q}\cdot\vec{J}^{(0)}(q) = -[\rho^{(0)}[1], V_{mom}[1/m^2] + V_{LS}[1/m^2]] - [\rho^{(0)}[1/m^2], V_{loc}[1]] \ , \quad (2.7)$$

where (0) means, as before, isoscalar. On the other hand, performing a gauge transformation in the last two terms of the potential (2.6), the resulting exchange current $\vec{J}_G$ satisfies [5],[6]

$$\vec{q}\cdot\vec{J}_G^{(0)}(q) = -[\rho^{(0)}[1], V_{mom}[1/m^2] + V_{LS}[1/m^2]] . \quad (2.8)$$

Since the last term in eq.(2.7) does not show up in eq.(2.8), we see that the method of minimal substitution in the potential cannot generate the full isoscalar exchange current. This was also noted a rather long time ago by Stichel and Werner [10], who, however, could develop a method in which this term is generated via a FWT-transformation of a two-body Dirac equation. In ref.[5] the calculation of magnetic moments was performed using the current $\vec{J}_G$, and for certain potential models a large quenching of the spin g-factor $g_s^{(0)}$ was found. For example, using the potential of Erkelenz et al.[11], the quenching found in ref.[5] for the nuclear matter case was -13%. We have reexamined this problem [12] and found that the part of the exchange current which gives rise to the last term in eq.(2.7) is very important: The quenching is turned into a 5% enhancement due to this term.

## 3. ISOSCALAR MAGNETIC MOMENTS IN A ONE-BOSON EXCHANGE MODEL

Having discussed some theoretical aspects of isoscalar magnetic moments in sect.2, we now proceed to an explicit calculation [12] based on the exchange current contributions shown in fig.3.

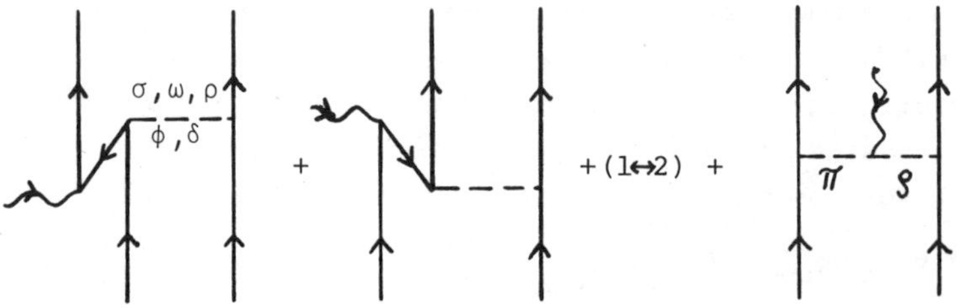


Fig.3 Isoscalar exchange currents considered in the present model.

The calculations are performed both for infinite nuclear matter and for finite nuclei with one particle or one hole outside an LS-closed core. For nuclear matter we will consider three cases which differ by the cut-off c characterizing the short-range correlation function $f(r)=\theta(r-c)$, and by the Fermi momentum $k_F$ in order to account for the fact that in finite nuclei the valence nucleon effectively feels a density less than the normal nuclear matter density. In the evaluation of the diagrams of fig.3 we use the one-boson exchange potential parameters of Holinde [13]. As discussed in the previous section, this exchange current satisfies the proper continuity equation (2.7). The numerical results are shown in table 1 in terms of deviations of the g-factors from their free values $g_s^{(0)}=0.88$, $g_l^{(0)}=0.5$ and $g_p^{(0)}=0$, i.e; we write for the correction of isoscalar magnetic moments

$$\delta\vec{\mu}^{(0)}=\delta g_s^{(0)}\vec{s} + \delta g_l^{(0)}\vec{l} + \delta g_p^{(0)}[Y^{(2)} x\ s^{(1)}]^{(1)} \qquad (3.1)$$

| | $\delta g_s^{(0)}$ | $\delta g_l^{(0)}$ | $\delta g_p^{(0)}/\sqrt{8\pi}$ |
|---|---|---|---|
| Matter c(fm) $k_F$(MeV) | | | |
| 0 260 | 0.081 | 0.025 | 0.007 |
| 0.7 260 | 0.023 | 0.036 | 0.010 |
| 0.7 206 | 0.012 | 0.018 | 0.006 |
| A=15 ($p_{1/2}^{-1}$) | 0.035 | 0.017 | 0.003 |
| A=17 ($d_{5/2}$) | 0.022 | 0.013 | 0.003 |
| A=39 ($d_{3/2}^{-1}$) | 0.037 | 0.018 | 0.004 |
| A=41 ($f_{7/2}$) | 0.026 | 0.014 | 0.003 |

Table 1 Deviations of g-factors due to exchange current processes.

We see that the nuclear matter estimates can qualitatively account for the corrections in finite nuclei, where both $g_s^{(0)}$ and $g_l^{(0)}$ are enhanced by about 3% to 4%. For comparison we note that for A=17 core polarization [1] quenches $g_s^{(0)}$ by about 13% and enhances $g_l^{(0)}$ by about 2%. Table 2 shows the resulting corrections for isoscalar magnetic moments and of the expectation value of the spin for finite nuclei. The exchange current contributions due to the diagrams of fig.3 are listed under $\delta\mu_{MEC}^{(0)}$, which we compare with the core polarization contributions $\delta\mu_{CP}^{(0)}$ calculated in ref.[1]. We also give the observed deviations from the Schmidt values $\delta\mu_{exp}^{(0)}$.

| | $\mu_{Schmidt}^{(0)}$ | $\delta\mu_{MEC}^{(0)}$ | $\delta\mu_{CP}^{(0)}$ | $\delta\mu_{exp}^{(0)}$ | $\langle s\rangle_{Schmidt}$ | $\langle s\rangle_{obs}$ | $\langle s\rangle_{calc}$ |
|---|---|---|---|---|---|---|---|
| A=15 | 0.187 | 0.003 | 0.040 | 0.031 | -0.167 | -0.094 | -0.061 |
| A=17 | 1.440 | 0.037 | -0.035 | -0.026 | 0.500 | 0.334 | 0.408 |
| A=39 | 0.636 | 0.019 | 0.063 | 0.070 | -0.300 | -0.164 | -0.134 |
| A=41 | 1.940 | 0.055 | -0.036 | -0.023 | 0.500 | 0.296 | 0.405 |

Table 2 Isoscalar magnetic moments and the expectation values of the spin.

In the presently used potential model the exchange current contributions seem to be rather large for particle states. However, we must note that the results are sensitive to the ratio of coupling constants $g_{\sigma NN}/g_{\omega NN}$, since generally the $\sigma$-meson gives rise to an enhancement and the $\omega$-meson to a quenching (comp. sect.2a). Thus, using models with larger $g_{\omega NN}$ would result in smaller corrections. The expectation values of the spin shown in table 2 are defined by

$$\mu^{(0)}_{exp} - \delta\mu^{(0)}_{MEC} = 0.38\langle s\rangle_{obs} + j/2 \qquad (3.2a)$$

$$\mu^{(0)}_{Schmidt} + \delta\mu^{(0)}_{CP} = 0.38\langle s\rangle_{calc} + j/2 \qquad (3.2b)$$

Note that if we would set $\delta\mu^{(0)}_{MEC}=0$ in eq.(3.2a), the resulting values of $\langle s\rangle_{obs}$ for the four cases shown in table 2 are -0.085, 0.432, -0.115 and 0.441, respectively, which are rather close to $\langle s\rangle_{calc}$. Therefore, again, the exchange current contributions seem to be somewhat too large. In any case, however, by comparing $\langle s\rangle_{Schmidt}$ with $\langle s\rangle_{obs}$ we see that the expectation value of the spin is strongly quenched, and second order core polarization is able to account for this quenching.

Finnaly, we note that the present exchange current calculation automatically includes both contributions shown in fig.2a and 2b, which arise by averaging the direct terms of the pair current operators shown in fig.3 over the nucleons in the Fermi sea. Thus, the cancellations discussed generally in sect. 2a are already taken into account in our numerical results. More explicitly, the contributions from the direct terms of the pair current operator due to $\sigma$-meson and $\omega$-meson exchange are for the nuclear matter case with c=0, $k_F$=260 MeV

$$\delta g^{(0)}_{1,dir} = -\frac{1}{2}\frac{\Sigma_s}{m} - \frac{1}{2}\frac{\Sigma_v}{m} = 0.163 - 0.120 = 0.043$$

$$\delta g^{(0)}_{s,dir} = -\frac{\Sigma_s}{m} - \frac{\Sigma_v}{m} = 0.327 - 0.240 = 0.087 \; .$$

This shows explicitly the cancellations between the two contributions fig.2a and 2b, and stresses once more the importance of vertex corrections.

The calculations which we described above refer to exchange

current contributions in lowest order. In nuclear matter we also have estimated the crossing terms between exchange currents and first order core polarization [12], and found that their contribution is negligibly small: The correction to $g_s^{(0)}$ is about one tenth of the values listed in table 1, and $g_l^{(0)}$ is left unchanged by the crossing terms.

## 4. CONCLUSION

In this work we have investigated two recent suggestions[3]-[6] which indicated appreciable exchange current contributions to isoscalar magnetic moments. On account of gauge invariance we found that in both treatments certain important terms seem to be omitted. We then performed explicit calculations using a one-boson exchange model for the exchange current operator. We found that the results are sensitive to the ratio of coupling constants $g_{\sigma NN}/g_{\omega NN}$. Due to this fact it is difficult to draw quantitative conclusions. In the present model calculation we found that both $g_s^{(0)}$ and $g_l^{(0)}$ are enhanced by about 3% to 4%, resulting in non-negligible corrections to isoscalar magnetic moments.

## References


1) K.Shimizu, M.Ichimura and A.Arima, Nucl.Phys.A226(1974)282
H.Hyuga, A.Arima and K.Shimizu, Nucl.Phys. A336 (1980) 363
2) I.S.Towner and F.C.Khanna, Nucl.Phys. A399 (1983) 334
3) A.Bouyssy, S.Marcos and J.F.Mathiot, Nucl.Phys. A415 (1984) 497
4) L.D.Miller, Ann.Phys. 91 (1975) 40
5) D.O.Riska, Phys.Script. 31 (1985) 107
6) D.O.Riska and M.Poppius, "Velocity Dependence and Nuclear Magnetic Structure", Univ. of Helsinki, preprint 1985
7) W.Bentz, A.Arima, H.Hyuga, K.Shimizu and K.Yazaki, Nucl.Phys. A436 (1985) 593
8) H.Kurasawa and T.Suzuki, Phys,Lett. 165B (1985) 234
9) W.Bentz, Nucl.Phys. A446 (1985) 678

10) P.Stichel and E.Werner, Nucl.Phys. A145 (1970) 257
11) K.Erkelenz, K.Holinde and K.Bleuler, Nucl.Phys. A139 (1969) 308
12) S.Ichii, W.Bentz and A.Arima, to be published
13) K.Holinde, Phys.Rep. 68 (1981) 122

# NUCLEAR ROTATION, SPONTANEOUS DEFORMATION AND HIGGS MECHANISM

Kazuo Fujikawa*) and Haruo Ui**)

*) Research Institute for Theoretical Physics, Hiroshima University, Takehara, Hiroshima 725

**) Department of Physics, Hiroshima University, Hiroshima 730

This paper is dedicated to Prof. G. Takeda for his 60th Birthday.

## §1. INTRODUCTION

In discussing nuclear structure in which rotational spectra appear very regularly, one usually starts by presupposing that the nucleus has a well-defined "intrinsic deformed" shape. Since the nucleus is an isolated system whose Lagrangian should be manifestly rotational invariant, one inevitably needs the concept of the "spontaneous break-down" of the rotation symmetry to rigorously define the concept of the "intrinsic deformation". Indeed, such a point of view has been emphasized by the very inventor of the concept of the "intrinsic deformation" — A. Bohr, who in his Nobel lecture[1)] deliberately stated;

*In a general theory of rotation, symmetry plays a central role. Indeed, the very occurrence of collective rotational degree of freedom may be said to originate in the breaking of rotational invariance, which introduces a "deformation" that makes it possible to specity an orientation of the system. Rotation represents the collective mode associated with such a spontaneous symmetry breaking (Goldstone boson).*

At this stage, we expect to have a degenerate ground-state rotation-band[2)], simply because the Nambu-Goldstone mode is massless. It is, therefore, necessary to go one step further in order to correctly understand the nuclear "massive" rotation with a finite moment of inertia. In this connection, we note that there is the well-known mechanism in the renormalizable gauge-field theory as well as in the theory of superconductivity — the Anderson-Higgs-Kibble mechanism, by which the massless Nambu-Goldstone mode is absorbed into the gauge field to acquire the mass. At first sight, since we have no gauge field in nuclear collective model, such a mechanism will not be applicable to nuclear rotation. We shall show in the present paper that the gauge field can be naturally introduced into the Bohr model by requiring that the Lagrangian be "form invariant" under a general time-dependent rotation. The obtained gauge field is identified with the angular velocity. We shall then show without any approximation that the massless Nambu-Goldstone mode arising from the spontaneous break-down of the rotation symmetry is absorbed into the gauge field to give rise to the massive rotation — i.e., the rotation motion with a finite moment of inertia. This typifies the Higgs mechanism[3] in its simplest version.

## §2. THE BOHR LAGRANGIAN AND ITS FIELD THEORETICAL INTERPRETATION.

Let us start by defining the field variables $\phi_m$ (m=2,1,0,-1,-2) which describe the nuclear quadrupole vibrations.[4] These variables are assumed to transform under a constant, time-independent rotation as

$$e^{-i\theta\cdot\hat{\mathbb{L}}}\,\phi(t)\,e^{i\theta\cdot\hat{\mathbb{L}}} = e^{-i\theta\cdot\mathbb{L}}\,\phi(t) \qquad (2\text{-}1)$$

where $\hat{\mathbb{L}}$ in the left-hand-side is the angular momentum operator, while $\mathbb{L}$ in the right-hand-side is its representation for L=2, $\phi$ being the column vector. The Lagrangian of the Bohr model[4] is then defined by

$$\mathcal{L} = \frac{1}{2}|\dot{\phi}|^2 - V(\phi) \ , \qquad (2\text{-}2)$$

where the dot denotes the time-derivative. We adopt as usual the reality condition[4] on $\phi_m, \phi^*_{-m} = (-)^m\phi_m$, so that $|\dot{\phi}|^2 = (\dot{\phi}_0)^2 + 2|\dot{\phi}_1|^2 + 2|\dot{\phi}_2|^2$ . Here, the potential $V(\phi)$ is taken to be rotation-invariant.

The most general rotation-invariant potential is a function of two independent invariants,

$$C_2 = |\phi|^2 = (\phi_0)^2 + 2|\phi_1|^2 + 2|\phi_2|^2 ,$$

$$C_3 = (\phi\times\phi\times\phi) = \sqrt{35/2}\ \Sigma_{m_1m_2m_3} (22m_1m_2|2m)(22mm_3|00)\phi_{m_1}\phi_{m_2}\phi_{m_3}$$

$$= \phi_0[6|\phi_2|^2 - 3|\phi_1|^2 - (\phi_0)^2] - 3\sqrt{3/2}[\phi_1\phi_1\phi_{-2} + \phi_{-1}\phi_{-1}\phi_2] ,$$

where $C_2$ and $C_3$ are, respectively, the quadratic and cubic Casimir operators[5)] of the $O(3)\times T_5$ group[5)] — i.e., the semi-direct product of the five-dimensional abelian group $T_5$, whose generators are $\phi_m$, and the three-dimensional rotation group O(3).

In order to keep a close analogy to the renormalizable field theory, we take the potential containing the fields $\phi_m$ up to the fourth order. Our potential is then represented as

$$V(\phi) = (\mu^2/2)|\phi|^2 + B(\phi\times\phi\times\phi) + C|\phi|^4 . \tag{2-3}$$

The constant C is assumed to be always positive to ensure the stability of the system. Our Lagrangian is now written as

$$\mathcal{L} = \frac{1}{2}|\dot{\phi}|^2 - \frac{\mu^2}{2}|\phi|^2 - B(\phi\times\phi\times\phi) - C|\phi|^4 . \tag{2-4}$$

Let us comment on this Lagrangian from two different view-points. First, in the field theoretical stand-point, this Lagrangian can be viewed as that of the (d=1) field theory — the field theory defined in the one-dimensional space-time, i.e., in time coordinate alone — in which the five (d=1) fields $\phi_m$ are coupled by the potential $V(\phi)$. In this point of view, the rotation-invariance of the system should formally be regarded as the invariance of the field with respect to the internal O(3) symmetry. It is to be noted that the transformation law of $\phi$ in (2-1) can be viewed as a global gauge transformation in our (d=1) field theory. This point of view will play an essential role to our generalization of the Bohr Lagrangian which are going to develop in subsequent sections.

Second, in the context of nuclear collective model, this Lagrangian describes the five coupled anharmonic oscillations of nuclear quadrupole vibrations. These oscillations are all stable when $(\mu^2/2)>0$ in (2-4) and the spontaneous break-down of the rotation-invariance may not take place in this case. On the other hand, when $(\mu^2/2)<0$, some of these oscillations always become unstable and the groud state of the system develops in such a way that the ground state expectation values of some of the fields — in our case $\phi_0$ — are non-vanishing. As a result, the fields $\phi_{\pm 1}$ become the Nambu-Goldstone modes[6)] associated with this spontaneous break-down of rotation symmetry. At this stage, since the Nambu-Goldstone mode is massless, one would naively expect to have the degenerate groun-state rotation-band. Before discussing this problem further, we shall generalize the Bohr Lagrangian by adopting the most general invariance principle.

## §3. LOCAL GAUGE INVARIANCE AND GENERALIZED BOHR LAGRANGIAN.

The Lagrangian (2-4) has been constructed so as to be invariant under a constant time-independent rotation, $\exp\{-i\theta\cdot\hat{\mathbb{L}}\}$. We now generalize the Bohr Lagrangian by imposing the most general invariance principle[7)] to our system — i.e., we require that the Lagrangian should be "form-invariant" under the general time-dependent rotation, $\exp\{-i\theta(t)\cdot\hat{\mathbf{L}}\}$.

It is to be noted that — in the context of the (d=1) field theory — the time-independent rotation as in (2-1) can be regarded as "global" gauge transformation associated with the internal O(3) symmetry, while our general time-dependent rotation can be viewed as its generalization to "local" gauge transformation.[7)]

The Lagrangian (2-5) is obviously not invariant under this time-dependent rotation, simply because the time-derivative of the fields, $\phi$, cannot homogeneously transform by this local transformation. As usual in any modern gauge field theory, the form invariant Lagrangian can be directly obtained by replacing the time-derivative $\partial_t$ to the covariant derivative $D_t$ such that $D_t\phi(t)$ and $\phi(t)$ obey the same transformation law;

$$\phi(t) \longrightarrow \phi'(t) = e^{-i\theta(t)\cdot\mathbb{L}}\ \phi(t) \tag{3-1a}$$

$$D_t\phi(t) \longrightarrow D_t'\phi'(t) = e^{-i\theta(t)\cdot\mathbb{L}}\ D_t\phi(t) \tag{3-1b}$$

From the above, we have $D_t = \partial_t - iB(t) = \partial_t - i\omega(t)\cdot\mathbb{L}$, where $B(t) = \omega(t)\cdot\mathbb{L}$ is the 5×5 matrix field which should transform as

$$B(t) \longrightarrow B'(t) = e^{i\theta(t)\cdot\mathbb{L}}\ B(t)e^{i\theta(t)\cdot\mathbb{L}} + (1/i)[\partial_t e^{-i\theta(t)\cdot\mathbb{L}}]\ e^{i\theta(t)\cdot\mathbb{L}}\ , \tag{3-2a}$$

i.e.,

$$\omega(t)\cdot\mathbb{L} \longrightarrow \omega'(t)\cdot\mathbb{L} = e^{-i\theta(t)\cdot\mathbb{L}}\ \omega(t)\cdot\mathbb{L}\ e^{i\theta(t)\cdot\mathbb{L}} + \dot{\theta}(t)\cdot\mathbb{L}. \tag{3-2b}$$

The infinitesmal transformation law of $\phi$ and $\omega$ is represented as

$$\phi \longrightarrow \phi' = (1 - i\delta\theta\cdot\mathbf{L})\phi \tag{3-3a}$$

and

$$\omega \longrightarrow \omega' = \omega - \omega \times \delta\theta + \delta\dot{\theta} \tag{3-3b}$$

which exemplify a local non-abelian gauge transformation.[7)]

We shall now show that our gauge field $\omega(t)$ introduced above can be identified as the angular velocity. For this purpose, we first note that our generalized Lagrangian to be constructed in the next section is "locally" gauge invariant. As a consequence of the local gauge invariance, the fields $\phi$ and $\omega$ do not all represent independent physical degree of freedom. From this redundancy, we can in principle choose a gauge among many possible choices. Now, consider the (time-like axial) gauge $\omega(t)=0$. With this choice of gauge, we obtain from (3-2) that $\omega'(t) = \dot{\theta}(t)$, from which we conclude that the gauge field can be identified as the angular velocity vector $\dot{\theta}(t)$.

## §4. KINETIC PART OF THE LAGRANGIAN AND CONSERVED CHARGES.

In terms of the covariant derivative, the kinetic part of our generalized Lagrangian is now given by

$$\mathcal{L}_{kin} = \frac{1}{2}|D_t\phi(t)|^2$$
$$= \frac{1}{2}|\dot{\phi}|^2 + \frac{1}{2i}\{\dot{\phi}^+(\omega\cdot\mathbb{L})\phi - \phi^+(\omega\cdot\mathbb{L})\dot{\phi}\} + \frac{1}{2}\phi^+(\omega\cdot\mathbb{L})(\omega\cdot\mathbb{L})\phi \ . \qquad (4\text{-}1)$$

Since the total Lagrangian $\mathcal{L} = \mathcal{L}_{kin} - V(\phi)$ is invariant under the general time-dependent rotation, there exist the conserved charges associated to this local symmetry. These Noether charges can be calculated[8] by the standard procedure to obtain $J_k = \partial\mathcal{L}/\partial\omega_k$:

$$J_k = \frac{1}{2i}\{\dot{\phi}^+L_k\phi - \phi^+L_k\dot{\phi}\} + \frac{1}{2}\phi^+\{L_k(\omega\cdot\mathbb{L}) + (\omega\cdot\mathbb{L})L_k\}\phi \ , \qquad (4\text{-}2)$$

which are, as expected, the components of the total angular momentum of the system. Indeed, it can be confirmed that $J_k$ satisfies the conventional commutation relations of the angular momentum when one quantizes $\phi$.

It is interesting to note that the relation $J_k = \partial\mathcal{L}/\partial\omega_k$, where $\omega_k$ is the angular velocity, has been well-known in the elementary theory of molecular vibration and rotation motion. Further analogy to the molecular motion can be highlightened when we write

$$\mathbf{J} = \mathbf{J}^{(vib)} + \mathbf{J}^{(rot)} \ , \qquad (4\text{-}3)$$

where

$$\mathbf{J}^{(vib)} = \frac{1}{2i}\{\dot{\phi}^+\mathbb{L}\phi - \phi^+\mathbb{L}\dot{\phi}\} \ ; \quad \mathbf{J}^{(rot)} = \frac{1}{2}\phi^+\{\mathbb{L}(\omega\cdot\mathbb{L}) + (\omega\cdot\mathbb{L})\mathbf{L}\}\phi.$$

In terms of $\mathbf{J}^{(vib)}$ and $\mathbf{J}^{(rot)}$, $\mathcal{L}_{kin}$ is represented as

$$\mathcal{L}_{kin} = \frac{1}{2}|\dot{\phi}|^2 + \omega\cdot\mathbf{J}^{(vib)} + \frac{1}{2}\,\omega\cdot\mathbf{J}^{(rot)}. \qquad (4\text{-}4)$$

Further, it is simple to show that $\mathbf{J}^{(rot)}$ can formally be written as $\mathbf{J}^{(rot)} = \mathcal{I}\omega$, so that the last term in (4-4) reads $\mathcal{L}^{(rot)} = \frac{1}{2}\omega\mathcal{I}\omega$, where the symmetric "dynamics" $\mathcal{I}$ is given by $\mathcal{I}_{k\ell} = \partial J_k/\partial\omega_\ell = \partial J_k^{rot}/\partial\omega_\ell = (1/2)\phi^+(L_kL_\ell + L_\ell L_k)\phi$. By comparing above formulas to those of the elementary theory of molecular rotation-vibration, one realizes that our gauge field $\boldsymbol{\omega}$ can be naturally identified with the angular velocity vector. Namely, in the context of the molecular physics, the first term in (4-4) combined with the potential term corresponds to the vibration, the second term the rotation-vibration coupling (Coliori term) and the last is the kinetic energy of the rotation.

In the context of the (d=1) field theory, our Lagrangian $\mathcal{L} = \mathcal{L}_{kin} - V(\phi)$, where $\mathcal{L}_{kin}$ is defined by (4-1) and $V(\phi)$ by (2-4), describes the system of the Higgs field $\phi$ coupled to the gauge field $\boldsymbol{\omega}$ in "locally" gauge-invariant manner. This Higgs field $\phi$ transforms as the five-dimensional representation under the internal O(3) symmetry group of the system and the suffix k of the gauge field $\omega_k$ refers to that of the internal O(3) space in just the same way as the SU(3) colour suffices of the gluon field in QCD. In this connection, we note that there is no field strength of our gauge field $\omega(t)$, because we are treating (d=1) gauge field theory. In other words, the free Lagrangian of the gauge field alone does not exist in the (d=1) gauge field in contrast to the (d≥2) gauge field.

## §5. POTENTIAL PART OF $\mathcal{L}$ WHEN $(\mu^2/2)<0$ — NAMBU-GOLDSTONE MODE.

We now treat the potential part of the Lagrangian in which $\mu^2$ is negative in (2-4). In this case, the quadrupole vibration becomes unstable and the field $\phi$ develops a finite "vacuum value": $\langle\phi\rangle_{\text{time average}} = 0$. Namely, the nucleus spontaneously deforms in a certain direction as will be shown shortly. We choose the direction of the deformation in the direction of $\phi_0$-component alone;

$$\phi_0 = v_0 + \phi_0', \quad \text{where} \quad \langle 0|\phi_0|0\rangle = v_0. \tag{5-1}$$

It can be proved[8] that this choice is sufficient to our specific form of the potential containing the $\phi$ up to the fourth order. Further, it can be proved[8] that our potential leads to an axial symmetric deformation, prolate or oblate depending upon whether $v_0$ is positive or negative.

We denote that $\phi' = (\phi_2,\phi_1,\phi_0',\phi_{-1},\phi_{-2})$ and $\phi = \phi' + v$. Introducing these into $V(\phi)$ in (2-4), we have

$$V(\phi'+v) = V(v) + \frac{\partial V(v)}{\partial v}\phi_0' + \frac{1}{2}\frac{\partial^2 V(v)}{\partial v^2}(\phi_0')^2 + \frac{1}{v}\frac{\partial V(v)}{\partial v}|\phi_1|^2$$

$$+ \left(\frac{1}{v}\frac{\partial V(v)}{\partial v} + 9vB\right)|\phi_2|^2 + 4Cv\phi_0'|\phi'|^2 + B(\phi'\times\phi'\times\phi') + C|\phi'|^4. \tag{5-2a}$$

where $$V(v) = v^2(\mu^2/2 - Bv + Cv^2)\ . \tag{5-2b}$$

From the above formula, one sees that the value of v can be determined in several equivalent ways; 1) The constant term $V(v)$ which is independent of any component of the field $\phi'$ should be minimum. 2) The term linear in $\phi_0'$ should vanish. 3) The coefficient of $|\phi_1|^2$ should vanish for a finite value of v because of the Nambu-Goldstone theorem.[6] 4) The $\phi_0'$ oscillation should be stable. We start from the first view-point that $V(v)$ be minimum at v. From the extremum condition of $V(v)$, $\partial V(v)/\partial v = v(\mu^2-3Bv+4Cv^2)=0$, we have $v=0$ and $v_\pm=(3B\pm\sqrt{9B^2-16\mu^2C})/8C$. Notice that $\mu^2<0$ and $C>0$, so that $9B^2-16\mu^2C$ is always positive.

It is now straightforward to show that $v=0$ gives the local maximum of $V(v)$, while $v_\pm$ gives us two local minima of $V(v)$. Further, the absolute minimum of $V(v)$ is given either by $v_+$ or $v_-$ depending upon whether B is positive or negative. In this absolute minimum of $V(v)$ at $v_0$, $V(\phi' + v_0)$ now reads

$$V(\phi' + v_0) = \frac{1}{3} v_0^2(\frac{\mu^2}{2} - Cv_0^2) + (-\frac{\mu^2}{2} + 2Cv_0^2)(\phi_0')^2 + 9v_0B|\phi_2|^2$$

$$+ 4Cv_0\phi_0'|\phi'|^2 + B(\phi'\times\phi'\times\phi') + C|\phi'|^4 \ . \tag{5-3a}$$

Here,
$$v_0 = \pm[3|B| + \sqrt{(3B)^2 - 16C\mu^2}]/8C, \tag{5-3b}$$
( + sign when $B>0$ and - sign when $B<0$ )

so that $v_0B$ always positive at this value of $v_0$.

From the above form of (5-3), one observes that (1) first term is always negative, (2) the coefficients of $(\phi_0')^2$ and $|\phi_2|^2$ are positive so that $\phi_0'$ and $\phi_2$ oscillations are always stable, ($B\neq 0$), and that (3) the $|\phi_1|^2$ term is missing. The last point is nothing but a consequence of Nambu-Goldstone theorem. Namely, when a continuous symmetry of the Lagrangian — in our case, the rotation symmetry which can be formally viewed as the internal O(3)-symmetry of the (d=1) field theory — is spontaneously broken such as in (5-1), there always appear the zero-energy modes corresponding to this broken symmetry. In the present case, the vacuum in (5-1) violates the rotation symmetry around the x and y axes so that we have the Nambu-Goldstone modes corresponding to $\phi_{\pm 1}$.

## §6. GAUGE-FIXING AND HIGGS MECHANISM.

Our generalized Lagrangian contains the eight variables $\phi$ and $\omega$ just to describe the five degrees of freedom originally specified by $\phi$. This means that we have to impose "three subsidiary conditions" to discuss the dynamics of the system. These subsidiary conditions are, of course, the "gauge condition" in conventional gauge theory.

The following gauge condition is commonly adopted when the spontaneous symmetry-breaking takes place. Unitary Gauge Condition[9)] — sometimes called the "physical gauge" — is specified by the vanishing Nambu-Goldstone modes. For our case, it is given by

$$\phi_{\pm 1}(t) = 0 \quad \text{and} \quad \omega_z(t) = 0. \tag{6-1}$$

With this choice of the gauge, the total angular momentum of our system is calculated from (4-2) and (4-3) by replacing $\phi$ with $v_0 + \phi'$. The result[8] is

$$J_z = (1/i)\{(\dot{\phi}_2^*\phi_2 - \phi_2^*\dot{\phi}_2) - (\dot{\phi}_{-2}^*\phi_{-2} - \phi_{-2}^*\dot{\phi}_{-2})\}$$

$$= (2/i)(\dot{\phi}_2^*\phi_2 - \phi_2^*\dot{\phi}_2) , \tag{6-2a}$$

$$J_+ = F\omega_+ + G\omega_- \quad \text{and} \quad J_- = G^*\omega_+ + F\omega_- , \tag{6-2b}$$

where

$$F = 2\{3(\phi_0)^2 + 2|\phi_2|^2) = 6v_0^2 + 12v_0\phi_0' + 6(\phi_0')^2 + 4|\phi_2|^2 ,$$

$$G = 4\sqrt{6}\phi_{-2}\phi_0 = 4\sqrt{6}(v_0\phi_2^* + \phi_2^*\phi_0') \ ; \quad G^* = 4\sqrt{6}(v_0\phi_2 + \phi_2\phi_0').$$

The kinetic part of the Lagrangian is given by

$$\mathcal{L}_{kin} = (1/2)(\dot{\phi}_0')^2 + |\dot{\phi}_2|^2 + \mathcal{L}_{kin}^{(coll)} , \tag{6-3}$$

where

$$\mathcal{L}_{kin}^{(coll)} = \frac{1}{2}G^*(\omega_+)^2 + F\omega_+\omega_- + \frac{1}{2}G(\omega_-)^2 = \frac{1}{2}(\omega_-J_+ + \omega_+J_-). \tag{6-4}$$

In order to rewrite the above formula in a familiar form of nuclear collective model, we use (6-2b) to express $\omega_\pm$ in terms of $J_\pm$;$\omega_+ = (FJ_+-GJ_-)/(F^2-|G|^2)$ and $\omega_- = (FJ_--G^*J_+)/(F^2-|G|^2)$. Introducing these results into (6-4), we finally obtain

$$\mathcal{L}_{kin}^{(coll)} = \frac{1}{2\mathcal{I}}(J^2 - J_z^2) - \frac{1}{4\mathcal{I}}\{(G^*/F)(J_+)^2 + (G/F)(J_-)^2\} , \tag{6-5}$$

where the (variable) moment of inertia $\mathcal{I}$ is defined by

$$\frac{1}{2\mathcal{I}} = \frac{F}{F^2 - |G|^2} = \frac{1}{2}\,\frac{3(v_0 + \phi_0')^2 + 2|\phi_2|2}{[3(v_0 + \phi_0')^2 - 2|\phi_2|^2]^2} . \tag{6-6}$$

We summarize below the exact formula of our total Lagrangian obtained by adopting the unitary gauge condition;

$$\mathcal{L} = \frac{1}{2}(\dot{\phi}_0')^2 + |\dot{\phi}_2|^2 + \frac{1}{2\mathcal{I}}(J^2 - J_z^2) - \frac{1}{2\mathcal{I}}\frac{\sqrt{6}(v_0 + \phi_0')}{3(v_0 + \phi_0')^2 + 2|\phi_2|^2}\{\phi_2(J_+)^2 + \phi_{-2}(J_-)^2\}$$

$$- \frac{1}{3}v_0^2(\frac{\mu^2}{2} - Cv_0^2) - (\frac{\mu^2}{2} + 2Cv_0^2)(\phi_0')^2 - 9v_0B|\phi_2|^2 - 4Cv_0\phi_0'|\phi'|^2$$

$$- B(\phi' \times \phi' \times \phi') - C|\phi'|^4 , \qquad (\mu^2/2 < 0 \quad \text{and} \quad C > 0) \qquad (6\text{-}7)$$

where the "vibration-dependent" moment of inertia $\mathcal{I}$ is defined by (6-6) and the value of $v_0$ is given by (5-3b). Here, $\phi'$ denotes $(\phi_2, 0, \phi_0', 0, \phi_{-2})$.

In short, what we have done in obtaining (6-7) is that we replaced the massless vibtators $\phi_{\pm 1}$ in (5-3) by the massive rotational variables $J_\pm$. This procedure in ordinary field theory is generally referred to as the "Higgs mechanism".

Details together with the discussions of the results in both the field theoretical and nuclear structure view-points can be found in ref.8.

It is our great pleasure to dedicate this paper to Professor G.Takeda on his 60th birthday.

## References


1) A. Bohr, Rev. Mod. Phys. 48 (1976) 305.
2) H. Ui and G. Takeda, Prog. Theor. Phys. 70 (1983) 176; See also G.Takeda and H. Ui, Prog. Theor. Phys. 69 (1983) 1146.
3) P.W. Higgs, Phys. Lett. 12 (1964) 132, 13 (1964) 508; Phys. Rev. 145 (1966) 1156; F. Englert and R. Brout, Phys. Rev. Lett. 13 (1964) 32; G.S. Guralnik, C.R. Hagen and T.W.B. Kibble, Phys. Rev. Lett. 13 (1964) 585; T.W.B. Kibble, Phys. Rev. 155 (1967) 1554; See also P.W. Anderson, Phys. Rev. 130 (1963) 439.
4) A. Bohr, K. Dan Vidensk Selsk. Mat. fys. 26 (1952) No.14; A. Bohr and B.R. Mottelson, ibid. 27 (1953) No.16, "Nuclear Structure" Vol.II (Benjamin, N.Y.) 1975.
5) H. Ui, Prog. Theor. Phys. 44 (1970) 153.
6) Y. Nambu and G. Jona-Lasinio, Phys. Rev. 122 (1961) 345;J. Goldstone, Nuov. Cim. 19 (1961) 154; J. Goldstone, A. Salam and S. Weinberg, Phys. Rev. 127 (1962) 965.
7) C.N. Yang and R.L. Mills, Phys. Rev. 96 (1954) 191.
8) K. Fujikawa and H. Ui, to appear in Prog. Theor. Phys.
9) S. Weinberg, Phys. Rev. D7 (1973) 1068; K. Fujikawa, B.W. Lee and A.I. Sanda, Phys. Rev. D6 (1972) 2923; E.S. Abers and B.W. Lee, Phys. Report C9 (1973) 1 and references therein.

# NUCLEAR GIANT QUADRUPOLE STATES

Toshio Suzuki
Research Institute for Fundamental Physics
Kyoto University
Kyoto 606, Japan

## §I INTRODUCTION

One of the most interesting collective states of nuclei is the giant quadrupole ($2^+$) resonance state which is observed throughout the periodic table.[1] Characteristic features of the giant resonance state are threefold. First, the excitation energy is given by $65/A^{1/3}$ MeV, where A denotes the mass number of the nucleus. Second, the strength for excitation of the collective state almost exhausts the model-independent sum rule value for photoexcitation. Third, the width of the resonance state is about 3 to 4 MeV. We should say that the width is unexpectedly narrow, since the resonance state exists at high level-density region. The purpose of this paper is to show that in assuming nuclear forces with zero-range, the above three features are explained self-consistently with use of the model-independent sum rules for the nuclear four-current[2] and Tomonaga theory for collective motions.[3] In the following section, we will derive the sum rules and in §III review briefly the Tomonaga theory. The structure of the giant resonance state will be discussed in §IV. The final section will be devoted to a summary of this paper.

## §II SUM RULES

The nuclear density and current operators in a single particle model are described as

$$\hat{\rho}(\vec{r}) = \sum_{k=1}^{A} \delta(\vec{r}-\vec{r}_k),$$

$$\hat{\vec{j}}(\vec{r}) = \frac{1}{2M} \sum_{k=1}^{A} \{\vec{p}_k, \delta(\vec{r}-\vec{r}_k)\},$$

where M stands for the nucleon mass. These operators satisfy the following model-independent sum rule,[2,4)]

$$\sum_n <0|\hat{\vec{j}}(\vec{r})|n><n|\hat{\rho}(\vec{r}\,')|0> = -\frac{i}{2M}\rho_0(r)\vec{\nabla}_{\vec{r}}\delta(\vec{r}-\vec{r}\,'), \qquad (1)$$

where $\rho_0(r)$ denotes the ground state density to be spherical,

$$\rho_0(r) = <0|\hat{\rho}(\vec{r})|0>,$$

and the suffix $\vec{r}$ of $\vec{\nabla}$ the differentiation with respect to $\vec{r}$. Provided that nuclear forces, V, commute with $\hat{\rho}(\vec{r})$, the continuity equation for the nuclear four-current is written as

$$\vec{\nabla}\cdot\hat{\vec{j}}(\vec{r}) = i[\hat{\rho}(\vec{r}), H], \qquad (2)$$

where H denotes the total Hamiltonian,

$$H = T + V, \qquad (3)$$

T being the kinetic part,

$$T = \sum_{k=1}^{A} \frac{\vec{p}_k^{\,2}}{2M} .$$

Applying eq.(2) to the sum rule, eq.(1), we have the density-density sum rule,[5)]

$$\sum_n \omega_n <0|\hat{\rho}(\vec{r})|n><n|\hat{\rho}(\vec{r}\,')|0> = -\frac{1}{2M}\vec{\nabla}_{\vec{r}}\cdot\{\rho_0(r)\vec{\nabla}_{\vec{r}}\delta(\vec{r}-\vec{r}\,')\}, \tag{4}$$

where $\omega_n$ stands for the excitation energy of the nuclear state, $|n>$. Since the density operator satisfies

$$\int\hat{\rho}(\vec{r})f(\vec{r})d\vec{r} = \sum_{k=1}^{A} f(\vec{r}_k)$$

for an arbitrary function, $f(\vec{r})$, the above two sum rules, eqs. (1) and (4), yield further sum rules,

$$\sum_n <0|\hat{\vec{j}}(\vec{r})|n><n|\sum_{k=1}^{A} f(\vec{r}_k)|0> = -\frac{i}{2M}\rho_0(r)\vec{\nabla}f(\vec{r}), \tag{5}$$

$$\sum_n \omega_n |<n|\sum_{k=1}^{A} f(\vec{r}_k)|0>|^2 = \frac{1}{2M}\int\rho_0(r)\vec{\nabla}f(\vec{r})\cdot\vec{\nabla}f(\vec{r})d\vec{r}, \tag{6}$$

which will be used later.

## §III TOMONAGA THEORY

When it is possible to describe the canonical variables for collective motions in terms of the nucleon coordinates and momenta, Tomonage theory[3)] provides us with the method to separate the Hamiltonian, eq.(3), into three parts,

$$H = H_{coll} + H_{coup} + H_{intr}, \tag{7}$$

where $H_{coll}$ and $H_{intr}$ describe the collective motions and other nuclear motions, respectively, while $H_{coup}$ induces the coupling between the collective and other modes. In denoting the collective coordinate and momentum by $\xi$ and $\pi$, respectively, the three terms in eq.(7) are written, up to 2nd order as for $\xi$ and $\pi$, as,

$$H_{coll} = \frac{\pi^2}{2I} + \frac{1}{2}(T_2+V_2)\xi^2, \tag{8}$$

$$H_{coup} = (T_1+V_1)\xi, \tag{9}$$

$$H_{intr} = (T_0+V_0). \tag{10}$$

Here I denotes the mass parameter for the collective motion to be determined with

$$[T - \frac{\pi^2}{2I}, \xi] = 0. \tag{11}$$

To, $T_1$ and $T_2$ are given by

$$T_0 = T - i\xi[\pi,T] - \frac{1}{2}\xi^2[\pi, [\pi,T]] \tag{12}$$

$$T_1 = i[\pi,T] + \xi[\pi, [\pi,T]] \tag{13}$$

$$T_2 = -[\pi, [\pi,T]], \tag{14}$$

while $V_0$, $V_1$ and $V_2$ are obtained from $T_0$, $T_1$ and $T_2$, respectively, by replacing T in the above expressions by V. Thus, once we obtain the collective variables, the Hamiltonian is uniquely separated into three parts.

## §IV GIANT QUADRUPOLE STATES

Collective Variables

In order to find the collective coordinate first, we use the sum rule, eq.(6). By setting $f(\vec{r})=2z^2-x^2-y^2$ in eq.(6), we have the sum rule for quadrupole photoexcitation,[6]

$$\sum_n \omega_n |\langle n| \sum_{k=1}^{A} (2z_k^2-x_k^2-y_k^2)|0\rangle|^2 = S \qquad (15)$$

where S denotes the sum rule value,

$$S = 4A\langle r^2\rangle/M,$$

$\langle r^2\rangle$ being the mean square radius of the nucleus,

$$\langle r^2\rangle = \frac{1}{A}\int \rho_0(r) r^2 d\vec{r}.$$

It is known experimentally[1] that the strength for the giant quadrupole state denoted from now on by $|\omega_2\rangle$ almost exhausts the sum rule value, S. Hence, eq.(15) is approximately written as,

$$\omega_2 |\langle \omega_2| \sum_{k=1}^{A} (2z_k^2-x_k^2-y_k^2)|0\rangle|^2 = S. \qquad (16)$$

Because of this fact, it is reasonable to choose the collective coordinate, $\xi$, for the quadrupole state to be

$$\xi = C\sum_{k=1}^{A}(2z_k^{\,2}-x_k^{\,2}-y_k^{\,2}) \tag{17}$$

with C being a constant.

The canonical momentum, on the other hand, is obtained on the basis of the sum rule for the current, eq.(5). If $|\omega_2\rangle$ exhausts the sum rule value, S, eqs. (5) and (16) yield the transition current;

$$\langle 0|\hat{\vec{j}}(\vec{r})|\omega_2\rangle = -\frac{i\omega_2}{2MS}\langle 0|\sum_{k=1}^{A}(2z_k^{\,2}-x_k^{\,2}-y_k^{\,2})|\omega_2\rangle$$

$$\times\ \rho_0(r)\vec{\nabla}(2z^2-x^2-y^2).$$

Defining the velocity field, $\vec{v}(\vec{r})$, by

$$\langle 0|\hat{\vec{j}}(\vec{r})|\omega_2\rangle = \rho_0(r)\vec{v}(\vec{r}),$$

the above equation provides us with the irrotational and incompressible field,

$$\vec{v}(\vec{r}) \propto \vec{\nabla}(2z^2-x^2-y^2)$$

which satisfies $\vec{\nabla}\times\vec{v}=0$ and $\vec{\nabla}\cdot\vec{v}=0$. As was shown in ref. 3), the canonical momentum which induces such a velocity field is written as

$$\pi = \sum_{k=1}^{A}(2z_k p_{zk}-x_k p_{xk}-y_k p_{yk}). \tag{18}$$

We have obtained $\xi$ and $\pi$ which are expressed in terms of the nucleon coordinates and momenta. Unfortunately, however, the above $\xi$ and $\pi$ do not satisfy the canonical commutation relation, but do

$$[\xi, \pi] = i\hat{O}, \tag{19}$$

where

$$\hat{O} = C \sum_{k=1}^{A} \{4r_k^2 - 2(2z_k^2 - x_k^2 - y_k^2)\}.$$

In order to have the canonical relation, therefore, we need an approximation. Let us replace in eq.(19) $\hat{O}$ by its ground state expectation value. In taking $C=1/4A\langle r^2\rangle$, then we have

$$[\xi, \pi] \approx i.$$

This approximation would be valid for heavy nuclei. Indeed, in using a harmonic oscillator potential model, the mean deviation is estimated to be

$$\{\langle 0|\hat{O}^2|0\rangle - \langle 0|\hat{O}|0\rangle^2\}^{1/2}/\langle 0|\hat{O}|0\rangle \approx 1.3/A^{2/3},$$

which is, for example, about 6% for A=100. Since this value is small enough for the present arguments, we will use $\xi$ in eq.(17) and $\pi$ in eq.(18) as the collective variables for the giant quadrupole states. We note, in fact, that the giant quadrupole state is not well defined as a collective state experimentally for light nuclei with A<40 [ref.1].

Mass Parameter

Once we obtain $\xi$ and $\pi$, the mass parameter for the collective motion is easily calculated according to eq.(11),

$$I = 2MA\langle r^2\rangle. \tag{20}$$

This is nothing but the mass parameter obtained in an irrotational and incompressible liquid drop model.[6]

Excitation Energy

The excitation energy of the collective state is given by eq.(8) as

$$\omega_2 = \{(T_2+V_2)/I\}^{1/2}. \tag{21}$$

The mass parameter, I, has already been obtained in eq.(20). The restoring force parameter, $(T_2+V_2)$, must be calculated according to eq.(14). The value of $T_2$ is easily obtained, neglecting the corrections of order $\sim A^{-2/3}$,

$$T_2 = 8\langle 0|T|0\rangle. \tag{22}$$

In order to estimate $V_2$, on the other hand, we have to assume nuclear forces. Let us assume nuclear forces to be provided by the zero-range forces,

$$V = V_0 \sum_{ij} \delta(\vec{r}_i-\vec{r}_j) + V_3 \sum_{ijk} \delta(\vec{r}_i-\vec{r}_j)\delta(\vec{r}_j-\vec{r}_k) + \cdots.$$

For this force, we can prove[7] that $[\pi,V]=0$, so that

$$V_2 = 0. \tag{23}$$

Eq.(23) reflects the fact that the potential-energy density given by the zero-range forces is a function of the nuclear density and, consequently, the volume conserving deformation in the incompressible mode does not change the potential energy. Because of $V_2=0$, the restoring force for the quadrupole mode stems only from the increase of the kinetic energy due to the deformation of the nucleus.

Eq.(21) with eqs.(20) and (22) yields

$$\omega_2 = \{4\langle 0|T|0\rangle/MA\langle r^2\rangle\}^{1/2}. \tag{24}$$

In using a harmonic oscillator potential model, the virial theorem gives the kinetic energy of the ground state as

$$\langle 0|T|0\rangle = M\omega^2 A\langle r^2\rangle/2, \tag{25}$$

where $\omega$ stands for the harmonic oscillator frequency and is determined so as to reproduce the nuclear radius,[6)]

$$\omega \approx 41/A^{1/3} \quad \text{(MeV)}.$$

Inserting eq.(25) into eq.(24), we finally obtain[6,8)]

$$\omega_2 = \sqrt{2}\omega \approx 58/A^{1/3} \quad \text{(MeV)},$$

which well explains the experimental value, $65/A^{1/3}$ MeV [ref.1].

Width

The width of the collective state comes from its coupling with other nuclear states through $H_{coup}$ in eq.(9). For the zero-range forces, we have $V_1=0$, because of $[\pi,V]=0$, so that there is no damping width due to the zero-range forces. On the other hand, $T_1$ is given by

$$T_1 = -\frac{1}{M}\sum_{k=1}^{A}(2p_{zk}^2-p_{xk}^2-p_{yk}^2) - \frac{2}{A\langle r^2\rangle}\langle 0|T|0\rangle\sum_{k=1}^{A}(2z_k{}^2-x_k{}^2-y_k{}^2), \tag{26}$$

neglecting $\sim A^{-2/3}$ corrections. This is a one-body operator which gives rise to the Landau damping of the collective state. In order to make the meaning of eq.(26) clear, let us use eq. (25) in a harmonic oscillator potential model. Then eq.(26) is described as

$$T_1 = -M\omega^2\sum_k\{(2z_k{}^2-x_k{}^2-y_k{}^2) + \frac{1}{M^2\omega^2}(2p_{zk}^2-p_{xk}^2-p_{yk}^2)\}. \tag{27}$$

In a harmonic oscillator potential model, the giant quadrupole state is composed of the particle-hole states where the particle state has the principal quantum number differing by 2 from that of the hole state. On the other hand, the operator, $T_1$, in eq. (27) excites particle-hole states only when the particle and hole have the same principal quantum number. Hence, $T_1$ can not produce the damping width of the giant resonance state. The observed narrow width is thus understood qualitatively. The width of 3 to 4 MeV would come from the neglected $A^{-2/3}$ corrections and the finite-range effect of nuclear forces.

## §V SUMMARY

In assuming nuclear forces with zero-range, the excitation energy, the narrow width and the collectivity of the giant quadrupole states are self-consistently understood on the basis of the sum rules for the nuclear four-current and Tomonaga theory for collective motions. It would be possible to apply the present method to the study of other giant resonance states like isoscalar monopole and isovector dipole states.[1] For those modes, one would have important contributions from $[\pi, V] \neq 0$ or $[\xi, V] \neq 0$ to the excitation energies and widths. In particular, $[\pi, V] \neq 0$ would produce the collision damping of those collective states. These are interesting problems to be investigated in future.

## References


1) J. Speth and A. van der Woude, Rep. Prog. Phys. 44 (1981) 719 and references therin.

2) T. Suzuki, Prog. Theor. Phys. 64 (1980) 1627.

3) S.I. Tomonaga, Prog. Theor. Phys. 13 (1955) 467; 482.

4) T. Suzuki and D.J. Rowe, Nucl. Phys. A286 (1977) 307.

5) H. Ui and T. Tsukamoto, Prog. Theor. Phys. 51 (1974) 1377 and references therein.

6) A. Bohr and B.R. Mottelson, Nuclear structure, vol.1 (Benjamin, New York, 1969); vol.2 (1975).

7) J. Martorell, O. Bohigas, S. Fallieros and A.M. Lane, Phys. Lett. B60 (1976) 313.

8) T. Suzuki, Nucl. Phys. A217 (1973) 182.

# The Vaporization of Hot Nuclei

H. Sagawa

Department of Physics, University of Tokyo
Bunkyoku, Hongo 7-3-1, Tokyo 113
JAPAN

[This paper is dedicated to Professor Gyo Takeda for his 60th anniversary]

## 1. Introduction

Recently, intermediate energy heavy-ion collisions have been studied intensively, looking for phase transitions under various thermodynamical circumstances. There are many open questions concerning the reaction mechanism during the early stages of the time evolution of hot nuclei which should be answered by studies based on microscopic theory. One of the fundamental questions is whether the global thermalization is really established or not in the initial stage of the head-on collision. Even if the nuclear matter is thermalized, the next question is whether this equilibrium lasts long enough to establish a first-order phase transition. Under these circumstances, it is not clear that the critical point behavior of the phase transition, originally discussed by M. E. Fisser [1], can be observed in thermalized nuclei created by heavy-ion collisions. However, in this note, I assume that the presence of two phases may be taken for granted.

The reaction which I would like to discuss is schematically shown in Fig.(1). I do not discuss how the global thermal equilibrium is established by heavy-ion reactions with typical bombarding energy $E_{Lab}$ =(10-100) MeV/A . This problem has been studied by statistical model calculations. A preliminary result of the nucleon-nucleus collision was reported by D. Boal [2], recently. We need to study a more elaborate simulation of the nucleus-nucleus collision with proper initial conditions in order to obtain a conclusive result on the creation of thermal equilibrium. In our model, the time evolution of the system will start from the instant that the compressed thermal nucleus has been created by the reaction.

## 2. The model

Let me now describe the model. The dynamical constraints on the system might be determined by physical situation shown in Fig.(1). First, the nucleus might expand or shrink depending on the initial conditions. This situation might be taken care of by a constraint $\lambda_1 r^2$ (or equivalently $\lambda_1 \rho(r)$). Secondly, the nucleus might move collectively in a certain direction due to its expansion. We need another constraint for this collective motion, namely, $\lambda_2'(\vec{p}.\vec{r}+\vec{r}.\vec{p})$. Thus, our model hamiltonian is described as a double-constrained one,

$$h''=h_{H-F}-\lambda_1 r^2 -\lambda_2(\vec{p}.\vec{r}+\vec{r}.\vec{p}) \qquad (1)$$

where $\lambda_1$ and $\lambda_2$ are independent lagrange multipliers.

The time evolution of the reaction process has been studied using two different approaches. The first is the mean field theory like TDHF [3], while the second is based on the classical equations of motion [4]. Temperature is easily corporated in the classical method, but there has been no serious calculation by TDHF at finite temperature, due to the numerical difficulty. In this paper, I would like to present a hybrid model which was recently proposed by G. F. Bertsch and the present author [5]. A basic assumption of the model is that global thermal equilibrium

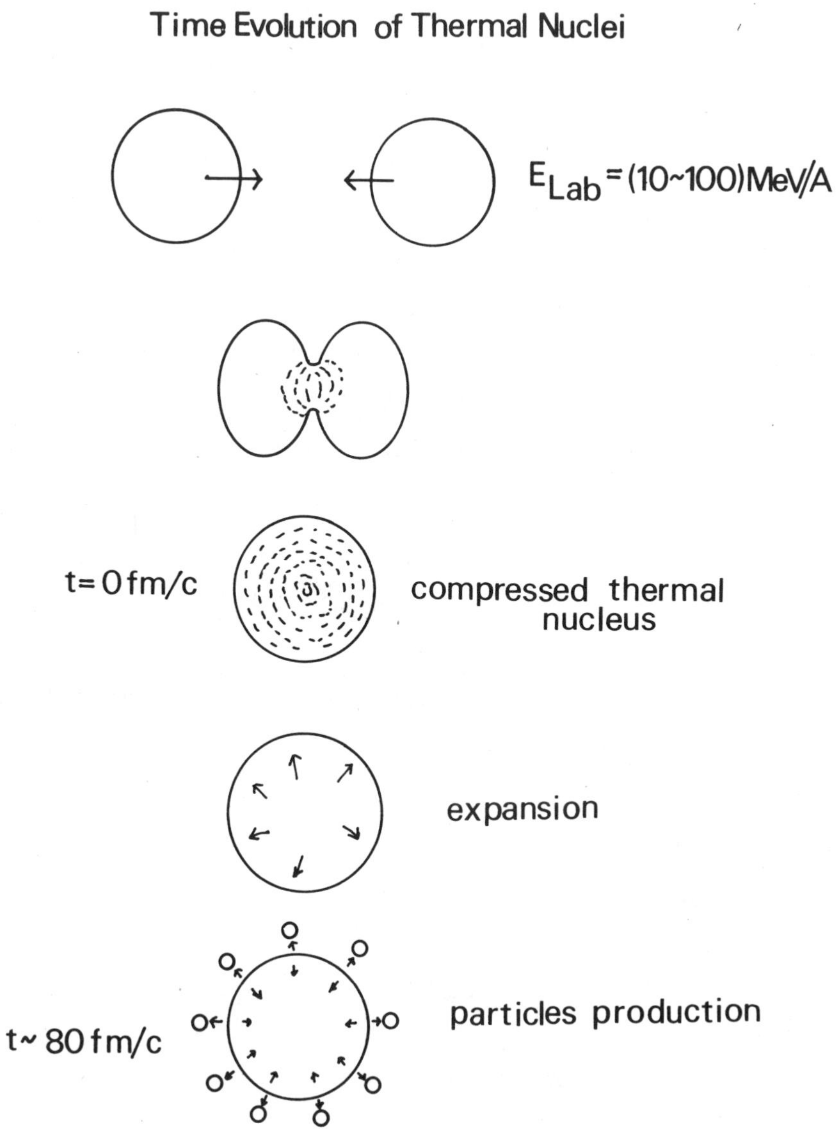
Time Evolution of Thermal Nuclei
E_Lab = (10~100) MeV/A
t = 0 fm/c
compressed thermal nucleus
expansion
t ~ 80 fm/c
particles production

Fig.1

is always existing during the expansion. This means that the proper time for the thermal equilibrium process is much shorter than that of the expansion process. Since statistical studies show two processes consuming almost same period of time, this assumption is not 100 % justified. However, it is worthwhile to pursue our model since there is no other study of the thermal system by quantum mechanical microscopic theory. Our final purpose is to calculate what fraction of the original nucleus will be left in the residual nucleus as a liquid phase after the expansion.

Hereafter I will summarize briefly some formulas of H-F theory at finite temperature. The variational method should apply for the thermodynamical potential

$$\Omega=\langle H\rangle-T.S-\mu\langle N\rangle \tag{2}$$

with respect to the grand canonical partition function instead of the hamiltonian $\langle H\rangle$, as in the zero-temperature case. The H-F equations to be solved are obtained as

$$h(\rho)\phi_i=\varepsilon_i\phi_i$$

$$\sum_i f_i(T)=\sum_i\{1+\exp[(\varepsilon_i-\mu)/T]\}^{-1}=A \tag{3}$$

where the density $\rho$ is defined by

$$\rho(r,T)=\mathrm{Tr}\{D\rho\}=\sum_i f_i(T)\phi_i^*\phi_i$$

The H-F equations are solved numerically in coordinate space using a Skyrme-type interaction SG II [6]. This parameter set SG II has a lower power of the density dependence $\rho^{1/6}$ in order to obtain a reasonable compression modulus $K_\infty$=218 MeV. The spin dependent Landau parameters are also reasonable compared with those of the G-matrix. At finite

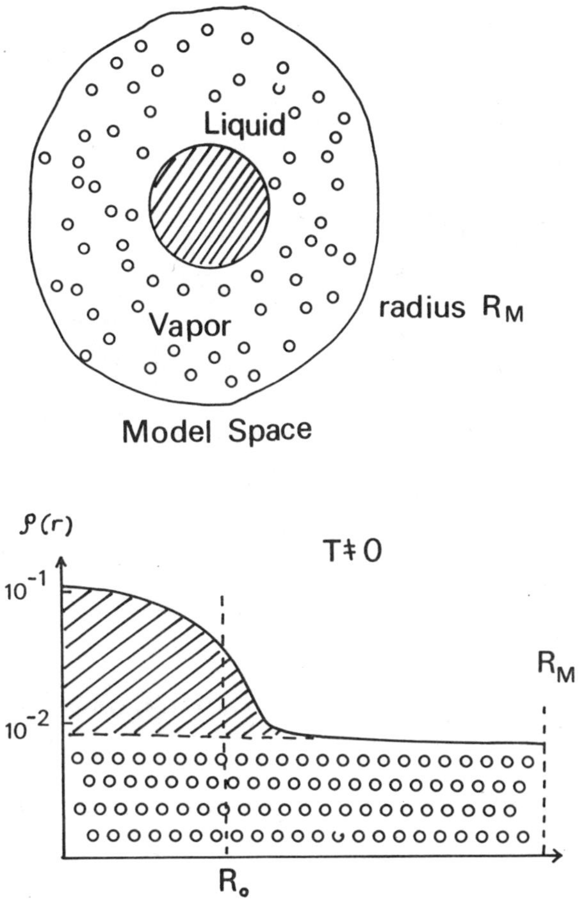


Fig.2

temperature, the H-F solutions have always some components spreading over the entire model space, besides those staying within the H-F potential, similar to the solutions at zero temperature. Certainly, the former component which is identified as a vapor phase is sensitive to the size of the model space $R_M$ in Fig.(2). This kind of solution might be physically meaningful in the following cases;

1) if one look at only the liquid solution(ref.(7)).

2) The static solution with the vapor phase might be qualitatively meaningful when one calculates the equation of state from the constrained H-F results, because the equation of state is rather stable for changes of the radius $R_M$ (ref.(8)).

3) The solution might be relevant for describing the time-evolution of the thermal nucleus with dynamical constraints (ref.(5)).

4) The size of the spherical box is determined by other physical constraints, for example, in the case of neutron stars [9].

I will now introduce the dynamical equations of motion based on the quantum mechanical commutators. The time dependences of the expectation values of the operators $r^2$ and $\vec{p}.\vec{r}$ are given by

$$id\langle r^2\rangle/dt=\langle[r^2,h_{H-F}]\rangle=\langle[r^2,h''+\lambda_1 r^2+\lambda_2\vec{p}.\vec{r}]\rangle$$

$$id\langle\vec{p}.\vec{r}\rangle/dt=\langle[\vec{p}.\vec{r},h_{H-F}]\rangle=\langle[\vec{p}.\vec{r},h''+\lambda_1 r^2+\lambda_2\vec{p}.\vec{r}]\rangle \qquad (4)$$

where the expectation value implies averaging over the grand canonical ensemble,

$$\langle O\rangle=\sum_i f_i(T)\langle\phi_i|O|\phi_i\rangle$$

$$h''\phi_i(\lambda_1,\lambda_2)=\varepsilon_i(\lambda_1,\lambda_2)\phi_i(\lambda_1,\lambda_2) \qquad (5)$$

Since the wave function $\phi_i$ is an eigenstate of the constrained hamiltonian, we can rewrite eq.(4) as simple dynamical equations

$$d\langle r^2\rangle/dt=2\lambda_2\langle r^2\rangle, \quad d\langle\vec{p}.\vec{r}\rangle/dt=-2\lambda_1\langle r^2\rangle \qquad (6)$$

These coupled equations describe essentially a harmonic vibration mode (the breathing mode as the first sound) in the small amplitude limit. The macroscopic model by H. Schultz et al.[10] also gives a similar equation for the harmonic vibration.

3. The result

The nucleus $^{40}Ca$ was taken for numerical study. We introduce 50 single-particle states as the neutron and proton configuration spaces, respectively. The vapor phase is extracted from another set of H-F solutions starting from zero potential with the same chemical potential $\mu$ as that of the solution with the vapor and liquid phases [7]. Since we are going to study the ratio of the vapor to the liquid phases in the thermalized nucleus, the chemical potential should be determined to keep the number of particles both in the liquid and the vapor fixed at A=40. We should notice that this prescription is different from that of Bonche et al. [7] who keep the number of particles only in the liquid phase fixed to be constant.

We now come to solve the time dependent equations (6) with physical initial conditions. We keep the excitation energy

$$E^{*}=\langle h_{H-F}\rangle^{T}_{\lambda_1,\lambda_2\neq 0}-\langle h_{H-F}\rangle^{T=0}_{\lambda_1=\lambda_2=0} \qquad (7)$$

to be constant. It is assumed that the system is compressed at t=0 fm/c to have a normal density $\rho(r=0)\sim 0.15\mathrm{fm}^{-3}$ at the center. Namely, the lagrange multiplier $\lambda_1$ is taken to be negative, while the other multiplier $\lambda_2$ is zero at t=0 fm/c. There are no constraints for the temperature and the entropy during the time evolution. The equations (7) are solved with the time mesh $\Delta t$=10 fm/c and several different excitation energies (different initial temperatures $T_I$). The density profiles of the vapor and the liquid are shown in Fig.(3) at the beginning and the end of the first vibration with the temperature $T_I$=6 MeV on the l.h.s. and 12 MeV on the r.h.s.. Some numerical results for the time-evolution of $^{40}$Ca are also listed in Table 1. The values $\lambda_1,\lambda_2$ and the temperature T are calculated at each time step to fulfill equations (7) and conservation of the total excitation energy. We can see that the temperature T is going down when the system is expanding. Nevertheless, the entropy S stays almost constant. Thus, in the initial stages of the expansion, the system behaves following the isentropic equation of state, but not the isothermal one.

In the case of $T_I$=6 MeV (equivalently, the excitation energy $E^{*}$=197 MeV and the entropy S/A=1.43), the vapor phase reaches 16% of the total nucleon number at the end of the expansion. On the other hand, the vapor phase occupies almost the entire space at t=80 fm/c in the case of $T_I$=12 MeV. This result implies that the whole system blows up at the very beginning of the expansion and never make a compound-like nucleus at high temperatures. On the other hand, the system might oscillate until it cools down to a certain temperature, and expel some vapor particles

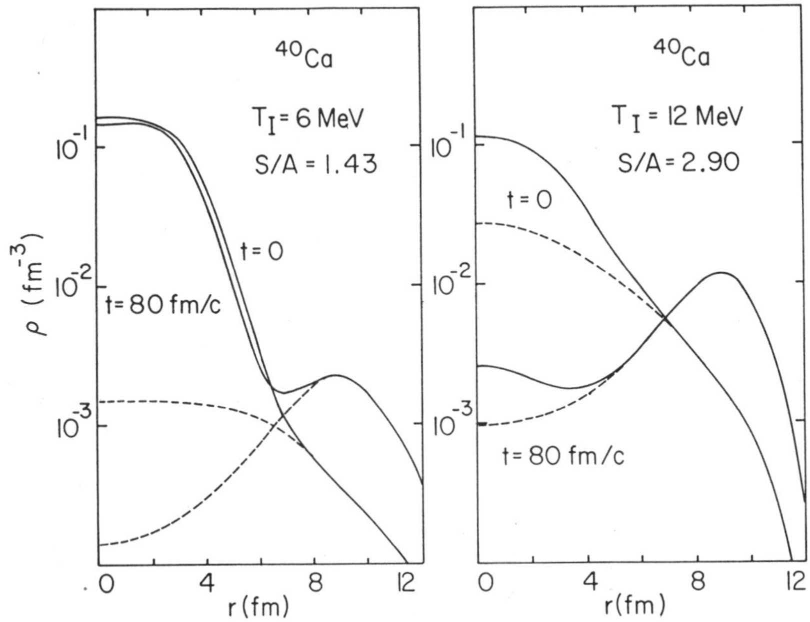

$^{40}$Ca
$T_I$ = 6 MeV
S/A = 1.43
t = 0
t = 80 fm/c
$\rho$ (fm$^{-3}$)
r (fm)
$^{40}$Ca
$T_I$ = 12 MeV
S/A = 2.90
t = 0
t = 80 fm/c
r(fm)


Fig.3

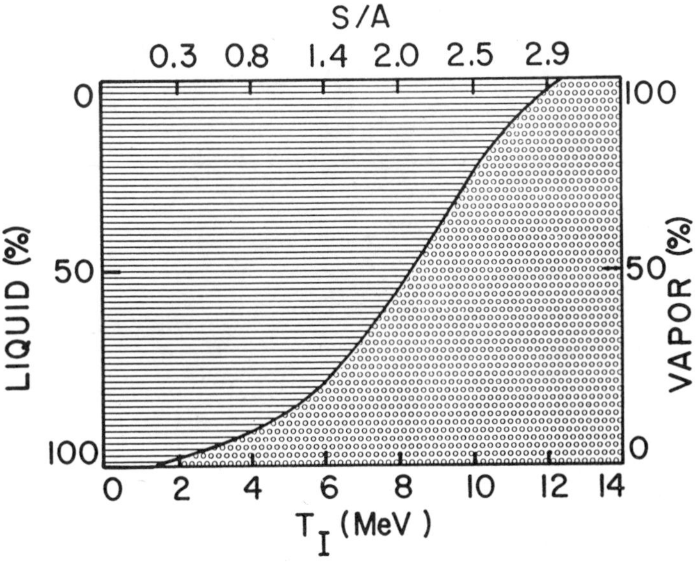

S/A
LIQUID (%)
VAPOR (%)
$T_I$ (MeV)


Fig.4

isotropically (dominantly single nucleons). In Fig.(4), the ratio of the vapor to the liquid phases is drawn for the different initial conditions. We can see that the vapor phase is less than 20% until the entropy S/A=1.5. The ratio of the vapor phase increases rapidly after this point, reaching to almost 100% at the entropy S/A=2.9. Thus, there is a certain critical entropy S/A for the change of the dominant phase from liquid to vapor. Our model prediction is very close to that of the macroscopic model. Some empirical evidence on the formation of the compound nucleus has been reported in ref. (11) at a temperature of around T=5 MeV.

Table 1
Time evolution of the hot nucleus $^{40}Ca$.

| $E^*$ (MeV) | $t$ (fm/$c$) | $\lambda_1$ (MeV/fm$^2$) | $\lambda_2$ (MeV/$\hbar$) | $r_m$ (fm) | $\langle p \cdot r \rangle$ ($\hbar$) | $T$ (MeV) | $S$ | $N_L$ (%) | $N_V$ (%) |
|---|---|---|---|---|---|---|---|---|---|
| 197 | 0 | −0.10 | 0.0 | 3.90 | 0.0 | 6.0 | 57 | 92.0 | 8.0 |
| | 20 | −0.03 | 0.78 | 4.07 | 0.31 | 5.5 | 58 | 89.0 | 11.0 |
| | 40 | 0.04 | 0.96 | 4.46 | 0.45 | 4.6 | 58 | 86.0 | 14.0 |
| | 60 | 0.09 | 0.52 | 4.84 | 0.29 | 4.0 | 59 | 82.0 | 18.0 |
| | 80 | 0.10 | −0.24 | 4.93 | −0.14 | 3.8 | 59 | 81.0 | 19.0 |
| | 100 | 0.06 | −1.36 | 4.57 | −0.62 | 4.2 | 59 | 84.0 | 16.0 |
| 597 | 0 | −0.20 | 0.0 | 5.21 | 0.0 | 12.0 | 116 | 44.0 | 56.0 |
| | 20 | −0.11 | 1.44 | 5.65 | 1.10 | 10.4 | 117 | 45.0 | 55.0 |
| | 40 | 0.01 | 1.7 | 6.80 | 1.85 | 7.3 | 117 | 38.0 | 62.0 |
| | 60 | 0.076 | 1.2 | 8.12 | 1.75 | 6.1 | 120 | 15.0 | 85.0 |
| | 80 | 0.11 | 0.37 | 8.90 | 0.7 | 5.9 | 121 | 3.0 | 97.0 |

Acknowledgements

The author would like to thank G. F. Bertsch and H. Toki for many enlightening discussions. He appreciates also D. W. L. Sprung for a careful reading of the manuscript.

References

1) M. E. Fisher, Physics 3 (1967) 256

C. B. Chitwood et al., Phys. Lett. 131B (1983) 289

2) D. Boal, preprint(1985)

3) A. Dhar and S. Das Gupta, Phys. Lett. 137B (1984) 303

J. Knoll and B. Strack, Phys. Lett. 149B (1984) 45

4) H. Jaqaman, A. Mekjian and L. Zamick, Phys. Rev. C29 (1984) 2067

A. Vicentini, G. Jacacci and V. Pandharipande, preprint (1984)

5) H. Sagawa and G. F. Bertsch, Phys. Lett. 155B (1985) 11

6) Nguyen van Giai and H. Sagawa, Phys. Lett. 106B (1981) 379

7) P. Bonche, S. Levit and D. Vautherin, Nucl. Phys. A427 (1984) 278

8) H. Sagawa and H. Toki, to be published

9) P. Bonche and D. Vautherin, Nucl. Phys. A372 (1981) 496

10) H. Schultz et al., Phys. Lett. 147B (1984) 17

11) S. Song et al., Phys. Lett. 130B (1983) 14

# THE RECENT DEVELOPMENT IN UNDERSTANDING THE PERIODIC TABLE OF ELEMENTS

Komajiro Niizeki
Department of Physics,
Tohoku University,
980 Sendai, Japan

Abstract: The recent development in understanding the periodic table of elements is reviewed. Our concern is focussed on the effects which make different elements of a group of the periodic table to have different chemical properties, which result in that different members of a homologous series of compounds have different physical properties. The most important effect is due to the effective repulsion of the valence orbital of an atom from the core region by orthogonality with the core orbitals with the same azimuthal quantum number.

## §1 Introduction

According to the modern theory of the periodic table of elements, a group of elements is characterized by the number of the valence electrons and the type of the valence shell which may consist of several subshells with different azimuthal quantum numbers. A set of all the elements with the same type of valence shells is referred to as a block. Each block is specified by the azimuthal quantum number of the subshell with the highest angular momentum. There are four blocks, i.e., s-, p-, d- and f-blocks, which have s-, sp-, sd- and sdf-valence shells, respectively. The s-block consists of Helium, alkalis (H, Li, Na, K, Rb and Cs) and alkaline earths (Ca, Sr and Ba). The p-block consists of other typical elements including the members of Zn group ( Be, Mg, Zn, Cd and Hg)[1]. The d-block consists of d-transition elements including noble metals (Cu, Ag and Au). The f-block consists of Lanthanoids and Actinoids. Each block of elements is divided into series by the difference in the principal quantum numbers of the relevant subshells. The p-block is divided into five series (np-series with n=2-6), the d-block into three series (nd-series with n=3-5) and the f-block into two series (4f- and 5f-series).

By the definition, all the elements of the same group of a periodic table have a common type of valence shells but belong to different series[2]. Similarities of the chemical properties among the members of a group is the basis of the periodic table. However, members of a single group can have diverse chemical properties[3]. It is one of the fundamental subjects of the chemistry or the condensed matter physics to understand what are the origins of making different members of a single group to have different chemical properties or different members of a homologous series of compounds to have different physical properties. Enormous efforts in the quantum theory of atoms, quantum chemistry and the theory of condensed matter physics have revealed several origins of the chemical diversity among the members of a group in the periodic table. It is the aim of the present paper to briefly report some of them.

In §2, we will discuss the origin of the first series singularities, i.e., the facts that elements of the first series in each block of elements have several exceptional properties from those of higher series. In §3, we will discuss other effects. We will present in §4 a few concluding remarks.

## §2. The first series singularities

The most important items which rule indivisual difference of each member of a group of elements are orbital energies ($\varepsilon_{n\ell}$ with n and $\ell$ being the principal and azimuthal quantum numbers) and spatial extensions ($\langle r\rangle_{n\ell}$) of the valence orbitals. Now, the number of nodes of a valence orbital specified by quantum numbers $n\ell$ is given by $n-\ell-1$, which represents also the number of core orbitals with the same azimuthal quantum number. The radial part of the orbital, $R_{n\ell}(r)$ is orthogonal to those of core orbitals, $R_{n'\ell}(r)$, $n'=\ell+1$, $\ell+2,\ldots,$ n. This orthogonality requirement strongly excludes the valence orbital from the core region. That is, the orthogonality requirement works effectively as a repulsive potential for the valence orbital. This can be best understood by the notion of the pseudo-potential[4]. The repulsion due to the core orbitals does not, however, work for a nodeless valence orbital, e.g., 1s-, 2p-, 3d- and 4f- valence orbitals. Consequently, such a valence orbital will be compact and has a large binding energy. Since the elements belonging to the first series of each block of elements have such valence orbitals, we may expect that these elements may have some exceptional properties in comparison with those of the elements belonging to higher series of the same block. We shall see this effect for each block of elements, separately.

i) The s-block

The nodeless orbital is 1s and two elements, H and He, have this orbital as valence orbitals. The ionization energy of H (13.6eV) is by far larger than those of other alkalis (the largest is 5.4eV for Li). This is the origin that among the members of alkali group only H is nonmental except under extremely high pressures. Furthermore, H never becomes a positive ion in a condensed phase, whereas other alkalis (except Li) are always positive ions in nomentallic condensed phases. H, instead, forms strong covalent bondings with non-metallic elements (particularly with B and C). It even becomes to a negative ion, $H^-$ in compounds with electro-positive elements such as alkalis or alkaline earths.

The fact that 1s orbital of H is very compact is the origin of the hydrogen bonding; the polarization of a hydrogen atom due to covalent bonding with an electronegative element such as N, O or F produces a strong electric field by which a negatively charged atom (usually, another N, O or F) is attracted towards the hydrogen atom. The hydrogen bonding is the origin of the fact that water is an excellent solvent. It plays, also, an essential role in the structures and the functions of DNAs, RNAs and proteins.

Liquid $^{4}He$ and liquid $^{3}He$ are quantum liquids, which do not freeze under an ambient pressure down to the absolute zero of temperature. This is not only because $^{4}He$ and $^{3}He$ have light masses but also because the intermolecular attraction of He is extremely weak, which is also a consequence of tightness of 1s orbital of He. In this respect, note that liquid hydrogen freezes under an ambient pressure though mass of $H_2$ is even lighter than $^{3}He$.

ii) p-block elements

The binding energy of the p-orbital of a 2p-series element is markedly strong in comparison with those of higher members of the same group. Moreover, the p-orbital has a spatial extension near that of the 2s-orbital, which is in marked contrast with the fact that the spatial extensions of p-ortibals of higher members are much larger than those of s-orbitals[5)]. The former is the origin of high electro-negativities of N, O and F. On the other hand, the latter weakens single bonds of C, N, O and F because of increased repulsion between lone pairs. This is the origin of the fact that C, N and O tend to form stable multiple bonds[5]. These are of vital importance in organic chemistry and biochemistry because C, N, and O together with H are the main constituents of organic compounds.

Uniqueness of Be, B or Ne among the members of Zn group, Boron group or noble gas group, respectively, is also well-known[3]. For example, the molecular diameter and the well depth parameter of Ne molecule are markedly smaller than those of the higher members (Ar, Kr and Xe).

The strong binding of 2p-orbital has some effects also on the properties of Li, which is taken to be an s-block element; Li can form a stable covalent bonding with C, which is in marked contrast with the fact that other alkalis ( from Na to Cs ) hardly form any covalent bonding in a nonmetallic condensed phase[3].

Uniqueness of the first series of elements among p-block elements is extensively discussed by Kutzelnigg[5], so that we will not discuss it any further.

iii) d-block elements

The orbital energies and the orbital extensions of the valence orbitals of d-block elements are shown in figs.1 and 2.[6] The gradual increase in the d-orbital binding energies and gradual shrinkage of the d-orbital extensions as functions of valency is due to the incomplete mutual screening of the nuclear charge among d-electrons.

It is evident from fig.1 that the binding energy of a

d-orbital of a 3d-element is considerably larger than those of the higher members of the same group. More striking is the compactness of the d-orbitals of 3d-elements in comparison with those of the higher members as is observed in fig.2.

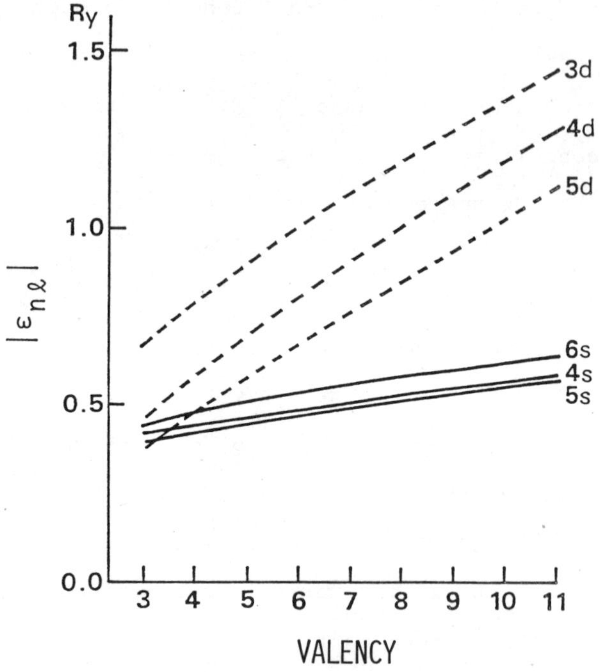


figure 1. The binding energies of the valence orbitals of d-block elements versus valency (the number of the valence electrons). The ordinate is scaled in Rydberg units (1Ry = 13.6eV).

The fact that the ratios of the orbital extensions of d-orbitals to those of s-orbitals are considerably small makes d-orbitals of d-block elements to have weaker abilities of forming chemical bondings than s-orbitals. This tendency is strongest for 3d-transition elements. This is the origin of the fact that many d-block transition elements, especially 3d-transition elements, have magnetic moments in simple substances and/or in metallic or nonmetallic compounds[8]; note that simply Cr, Mn, Fe, Co and Ni are, among d-block transition elements, anti-ferromagnetic or ferromagnetic in simple substances. We can also understand by the same reason that a higher series transition element is apt to have higher oxidation numbers in comparison with the first series member of the same group[3].

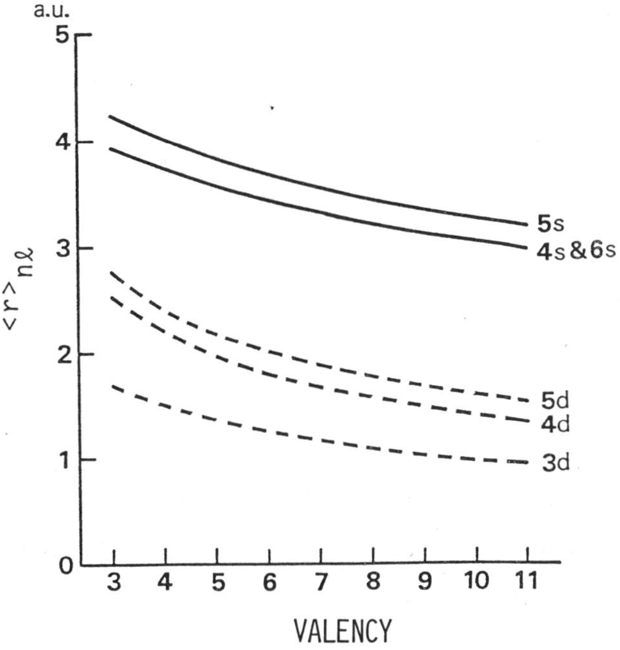


figure 2. The orbital extensions of the valence orbitals of d-block elements versus valency ( 1a.u. = Bohr radius = .529 × $10^{-10}$ m ). The curves for the 4s-orbital and the 6s-orbital can hardly be distinguished from each other in this scale.

iv) f-block elements

The f-orbitals of 4f-transition elements are so strongly localized ( $\langle r \rangle_{4f} \lesssim 1.1$ a.u. )[7] that they can hardly join in the chemical bonding in simple substances and compounds. This is the reason why the chemical properties of 4f-transition elements (the Lanthanoids) are so similar. The spatial extension of a 4f-orbital shrinks as a function of the atomic number by the same reason as that for a similar fact observed in the case of d-block elements. This is the origin of the Lanthanoid contraction, i.e., an empirical rule that an ionic radius of a trivalent Lanthanoid ion is a decreasing function of the atomic number[3].

The f-orbitals of earlier members of 5f-series elements are much more extended than those of 4f-transition elements ( $\langle r \rangle_{5f} \gtrsim 1.35$ a.u. up to Pu ) but those of other members are not necessary so because of the shrinkage of f-orbitals. ( $\langle r \rangle_{5f} \simeq 1$.a.u. for No )[7]. This makes f-orbitals of early 5f-transition elements to be chemically active but does not for later ones. Consequently, chemical properties and magnetic properties of earlier members of 5f-transition elements are considerably different from the corresponding members of 4f-transition elements[3].

## §3 The nonmonotonous change in the chemical and/or physical properties in a group of elements

i) p-block elements

The binding energies of s-valence orbitals of p-block elements are shown in fig.3.[9] The binding energy is roughly a decreasing function of the series number in a single group of elements. The decreasing trend is, however, inverted or weakened between the second series and the third series. Moreover, the trend is always inverted between the fourth and the fifth series. The nonmonotonous change of the binding

figure 3.

The binding energies of the s-valence orbitals versus the series number for each group of the p-block elements.

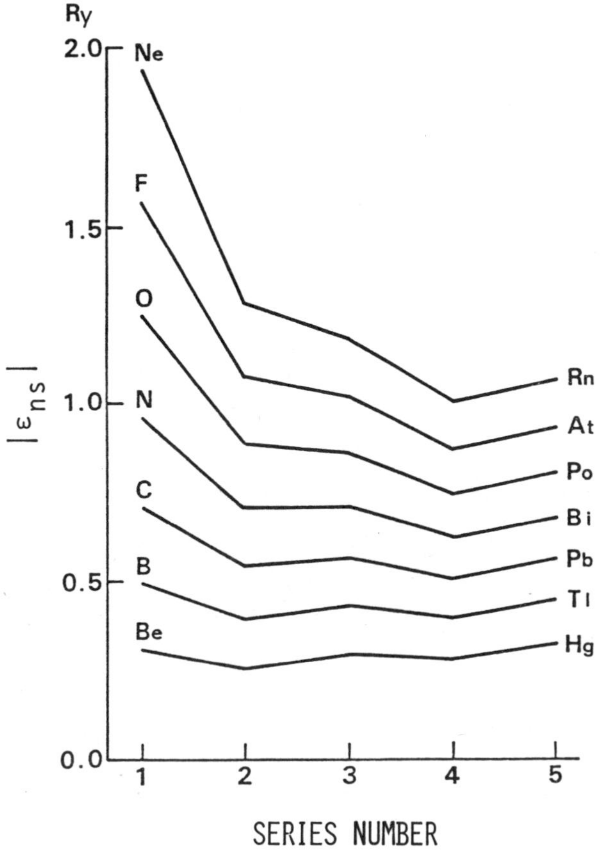


energies of s-orbitals among the members of a group of p-block elements gives rise to a nonmonotonous change of their chemical properties and/or of the physical properties of compounds containing the elements. We shall investigate below the origins of the nonmonotonous change.

We first consider the anomaly arising in the shift from the second series to the third one. An obvious difference between the two series lies in the types of the outermost shells of the cores; the electronic configuration of the outermost core shell of a second series atom or a third series atom is $2p^6$ or $3d^{10}$, respectively. The spatial extension of the 3d-orbitals of a third series atom is larger on account of the increased centrifugal barrier than that of the the 2p-orbitals of the second series atom of the same group. Consequently, the screening of the nuclear charge by the 3d-orbitals for the 4s-valence electrons is weaker than that by the 2p-orbitals for the 3s-valence electrons. This is the reason why the binding energies of the 4s-orbitals of earlier members of the third series are larger than those of the 3s-orbitals of the corresponding members of the second series.

This is not the case for the later members of the two series because the spatial extension of the 2s-orbital in the core of a second series element decreases faster as a function of the atomic number than that of the 3s-orbital of a third series element; the repulsive effect of the 2s-orbital of a later member of the second series elements on the 3s-valence orbital is weaker than that of 3s-orbital of the corresponding member of the third series on the 4s-valence orbital.

The electronic configurations of the outermost shells of the cores of the third and the fourth series are similar ($3d^{10}$ and $4d^{10}$, respectively). This results in the normal change in the binding energies of s-valence orbitals on shifting from the third to the fourth series of the p-block elements. On the other hand, the increase in the binding energies of s-valence orbitals on shifting from the fourth to the fifth series can not be ascribed to the presence of the 4f-orbitals in the cores of the fifth series elements because the outermost shell of the cores of the fifth series elements is not the 4f-shell but the 5d-shell[7]. The anomaly in fact originates from the relativistic effect[10),11)].

ii) d-block elements

It is well known that the chemical properties of a member of 4d-series elements are very similar to those of the corresponding member of the 5d-series. This originates from the similarity of the orbital energies and orbital extensions of s- and d-orbitals between the two series as seen in figs.1 and 2. It can be shown that the similarity would be spectacular but for the relativistic effect[7]. The relativistic effect is more important on 5d-series than on 4d-series; its effect on the 5d-series is to increase the binding energies of the 6s-orbitals, which indirectly decreases the binding energies of the 5d-orbitals through the enhanced screening of the nuclear charge due to the shrinkage of the 6s-orbitals. The fact that the decrease in the binding energies of d-orbitals on shifting from the 4d-series to the 5d-series is suppressed in the nonrelativistic calculation is derived from the weakness of the screening of the nuclear charge by the 4f-orbitals for the 5d-orbitals of a 5d-series element.

The relativistic effect has a significant influence on the chemical properties of the elements of the 5d-series[10),11)] (and also of the 5f-series[10)]).

## §4 Concluding remarks

An atom is a many electron system; the relevant forces are coulombic and the dynamics of the system is quantum mechanics. Its structure is determined as a consequence of a subtle balance among wave-mechanical effects and many body effects. Since the range of the coulomb potential is infinite, the system has no dimensionless parameters; the relativistic effect characterized by the fine structure constant is a minor effect except for heaviest elements and the nuclear mass can be taken to be infinite on the scale of electron mass. Therefore, there is no room for choice of the electronic properties of elements; the Creator has no parameters available for finely tuning the chemical properties of elements[12)]. Hence, it was lucky for him that the elements in our universe have such chemical properties as the living organisms can evolve!

## References and notes:

1) It is usual to include Be and Mg in alkaline earth group but we have adopted a different allotment.

2) The noble gas group includes an s-block element (He) in addition to p-block elements (from Ne to Xe).

3) F.A. Cotton and G. Wilkinson, 'Advanced Inorganic Chemistry', 3rd ed., John Wiley & Sons, Inc., New York 1972.

4) J.C. Phillips and C. Kleinman, Phys. Rev. 116, 287 (1959).
V. Heine, Solid State Physics, 24, 1 (1970).

5) W. Kutzelnigg, Angew. Chem. 23, 272 (1984).

6) The source of the data is the Dirac-Fock calculations by J.P. Desclaux[7)]. The spin-orbit splitting in the binding energies or in the orbital extensions is averaged over the multiplets. The electronic configurations assumed are $nd^{v-2}(n+1)s^2$ (n=3-5, v=3-11), though the ground configurations of some elements (especially a majority of 4d-series) are different from those. The missing data in ref.7) which compiles only the data for the ground state configurations are supplemented by an interpolation and/or an extrapolation.

7) J.P. Desclaux, At. Data. Nucl. Data Tables 12, 311 (1973).

8) J.B. Goodenough, 'Magnetism and the Chemical Bonding', Interscience publ., New York 1963.

9) The source of the data is ref. 7).

10) P. Pyykö and J.P. Desclaux, Acc. Chem. Res. 12, 276 (1979).

11) K.S. Pitzer, Acc. Chem. Res. 12, 271 (1979).

12) For 'fine-tuning', see the following article.

13) P.C. Davies, 'The Accidental Universe', Cambridge University press, Cambridge, 1982.

# III. GENERAL PHYSICS

# SOLUTIONS OF DISCRETE MECHANICS NEAR THE CONTINUUM LIMIT

T. D. Lee

Columbia University, New York, N. Y. 10027

Abstract

The time-distribution in classical discrete mechanics near the continuum limit is given for a non-relativistic particle moving in an arbitrary potential.

This research was supported in part by the U.S. Department of Energy.

# I. CLASSICAL DISCRETE MECHANICS (REVIEW)

Take the simplest example of a one-space-dimensional non-relativistic particle of unit mass moving in a potential $V(x)$. In the usual continuum mechanics the action is

$$A(x(t)) = \int_0^T \left[\tfrac{1}{2}\dot{x}^2 - V(x)\right] dt , \tag{1.1}$$

where $x(t)$ can be any smooth function of the time $t$. Keeping fixed the initial and final positions, say $x_0$ and $x_f$, at $t = 0$ and $T$, we determine the orbit of the particle by the stationary condition

$$\frac{\delta A}{\delta(x(t))} = 0 \tag{1.2}$$

which leads to Newton's equation

$$\frac{d^2X}{dt^2} = -\frac{dV}{dx} . \tag{1.3}$$

In the above, $x$ is the dynamical variable and $t$ is merely a parameter. Next, we review how this customary approach may be modified in the discrete version.[1]

Let the initial and final positions of the particle be the same

$$x_0 \quad \text{at} \quad t = 0 \qquad \text{and} \qquad x_f \quad \text{at} \quad t = T . \tag{1.4}$$

In the discrete mechanics we restrict the usual smooth path $x(t)$ to a "discrete path" $x_D(t)$, which is continuous but piecewise linear, characterized by $N$ vertices (as shown in Figure 1). In 1(a) we have the usual smooth path $x(t)$ of a non-relativistic particle in classical mechanics. Moving along $x(t)$ from $t = 0$ to $T > 0$, the time $t$ increases monotonically; this property is retained under the constraint restricting $x(t)$ to $x_D(t)$. Thus, as in Figure 1(b), we may label the $N$ vertices of $x_D(t)$

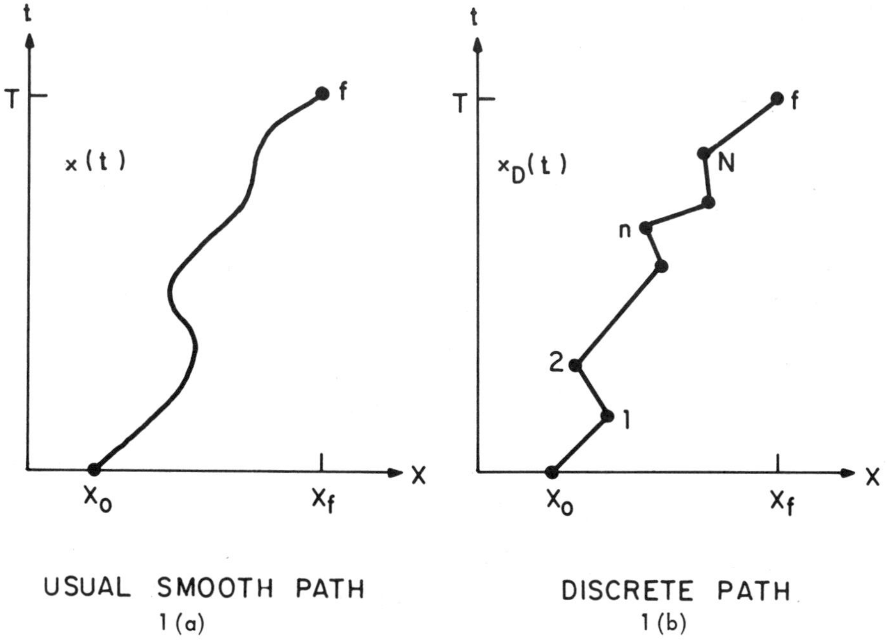


FIGURE 1.

consecutively as $n = 1,2, \cdots, N$, each of which carries a space-time position $x_n$ and $t_n$ with

$$0 < t_1 < t_2 < t_3 < \cdots < t_N < T . \qquad (1.5)$$

The nearest neighboring vertices are linked by straight lines, forming the discrete path $x_D(t)$, which also appears as a one-dimensional lattice with $n$ as lattice sites. In Figure 1(b), a variation of the space-time positions of these vertices changes the discrete path $x_D(t)$. However, a mere exchange of any two vertices clearly defines the same $x_D(t)$. This is because only the discrete path with unlabeled vertices has a physical meaning. There is no "individual" identity of any of the vertices. Thus, the time-ordered sequence (1.5) is not an additional restriction, but one that arises naturally when we pass from the usual continuum path $x(t)$ to the discrete one $x_D(t)$.

In the following, we shall keep the site-density

$$\frac{N+1}{T} \equiv \frac{1}{\epsilon} \tag{1.6}$$

fixed, and regard $\epsilon$ as a fundamental constant of the theory. The original continuum action integral (1.1) evaluated on such a discrete path $x_D(t)$ is the new discrete action[2]

$$A_D = A(x_D(t)) = \sum_n \left[\frac{1}{2}\frac{(x_n - x_{n-1})^2}{t_n - t_{n-1}} - (t_n - t_{n-1})\,\bar{V}(n)\right] \tag{1.7}$$

where

$$\bar{V}(n) = \frac{1}{x_n - x_{n-1}} \int_{x_{n-1}}^{x_n} V(x)\, dx \tag{1.8}$$

is the average of $V(x)$ along the straight line between $x_{n-1}$ and $x_n$.

Because the path $x_D(t)$ is completely specified by its vertices $n(x_n, t_n)$, a variation in $x_D(t)$ is equivalent to a variation in all the positions of its vertices

$$d\,[x_D(t)] = \prod_n [dx_n]\,[dt_n]\ . \tag{1.9}$$

Correspondingly, the dynamical equation (1.2) becomes the difference equations

$$\frac{\partial A_D}{\partial x_n} = 0 \tag{1.10}$$

and

$$\frac{\partial A_D}{\partial t_n} = 0\ . \tag{1.11}$$

We see that in this new mechanics the roles of $x_n$ and $t_n$ are quite similar. Both appear as dynamical variables. For each $x_n$ or $t_n$ we have one difference equation, (1.10) or (1.11). The former gives Newton's law on the lattice and the latter gives the conservation of energy

$$E_n \equiv \frac{1}{2}\left(\frac{x_n - x_{n-1}}{t_n - t_{n-1}}\right)^2 + \overline{V}(n) = E_{n+1} \quad . \tag{1.12}$$

In the usual continuum mechanics, conservation of energy is a consequence of Newton's equation. Here, these two equations (1.10) and (1.11) are independent. Altogether there are 2N such equations, matching in number the 2N unknowns $x_n$ and $t_n$ in the problem. Because the action $A_D$ is stationary under a variation in $x_n$ and in $t_n$ for every $n$, the discrete theory retains the translational invariance of both space and time, and that leaves the conservation laws of energy and momentum intact.*

Example 1

For a free particle $V(x) = 0$, (1.10) and (1.11) become degenerate; both give

$$v_n = \frac{x_n - x_{n-1}}{t_n - t_{n-1}} = \text{constant} . \tag{1.13}$$

The corresponding trajectory is a straight line, the same as in the continuum case.

Example 2

When $V(X) = gx$ with $g$ a constant, the solution of (1.10) and (1.11) can be readily found. We find in this case the spacing between successive $t_n$ to be independent of $n$ :

$$t_n - t_{n-1} = \epsilon \quad . \tag{1.14}$$

Correspondingly, $t_n = t_0 + n\epsilon$ and

$$x_n = x_0 + n v_1 \epsilon - \tfrac{1}{2} n(n-1)\, g \epsilon^2 \tag{1.15}$$

where $v_1$ is the initial velocity $(x_1 - x_0)/(t_1 - t_0)$ .

---

* Here, conservation of momentum means that the change of particle momentum is equal to the "impulse" generated by the potential.

For a more complicated $V(x)$, the distribution of $t_n$ is, in general, not uniform, as we shall see.

## II. NEAR THE CONTINUUM LIMIT

Consider now an arbitrary potential $V(x)$. Define

$$\epsilon_n \equiv t_n - t_{n-1} , \tag{2.1}$$

$$\ell_n \equiv x_n - x_{n-1} \tag{2.2}$$

and

$$\tau \equiv nT/(N+1) = n\epsilon \tag{2.3}$$

where, as in (1.6), $\epsilon = T/(N+1)$. When $n$ runs from 0 to $N+1$, $\tau$ varies from 0 to $T$. To obtain the continuum limit, we keep $T$ fixed but let $\epsilon \to 0$, and therefore $N \to \infty$. In this limit we regard

$$x_n = x(\tau) , \qquad t_n = t(\tau) \tag{2.4}$$

as $O(1)$, $\ell_n$ and $\epsilon_n$ as $O(\epsilon)$, but $\ell_n - \ell_{n-1} \cong \epsilon\, d\ell_n/d\tau$ as $O(\epsilon^2)$, etc. In terms of

$$V'(x) \equiv dV/dx , \quad V''(x) \equiv d^2V/dx^2 , \quad V'''(x) \equiv d^3V/dx^3 , \cdots ,$$
$$V_n \equiv V(x_n) , \quad V'_n \equiv V'(x_n) , \quad V''_n \equiv V''(x_n) , \quad V'''_n \equiv V'''_n(x_n) , \cdots , \tag{2.5}$$

we may expand (1.8) in powers of $\ell_n$ :

$$\overline{V}(n) = V_{n-1} + \frac{1}{2}\ell_n V'_{n-1} + \frac{1}{3!}\ell_n^{\,2} V''_{n-1} + \cdots . \tag{2.6}$$

Correspondingly (1.10), $\partial A_D/\partial x_n = 0$, becomes

$$v_n = v_{n-1} - \frac{1}{2}(\epsilon_n + \epsilon_{n-1}) V'_{n-1} - \frac{1}{3!}(\epsilon_n \ell_n - \epsilon_{n-1}\ell_{n-1}) V''_{n-1}$$
$$- \frac{1}{4!}(\epsilon_n \ell_n^{\,2} + \epsilon_{n-1}\ell_{n-1}^{\,2}) V'''_{n-1} + O(\epsilon^5) \tag{2.7}$$

where $v_n = (x_n - x_{n-1})/(t_n - t_{n-1}) = \ell_n/\epsilon_n$. Likewise (1.11), $\partial A_D/\partial t_n = 0$, gives

$$v_n^2 = v_{n-1}^2 - (\ell_n + \ell_{n-1}) V'_{n-1} - \frac{1}{3}(\ell_n^2 - \ell_{n-1}^2) V''_{n-1}$$

$$- \frac{1}{12}(\ell_n^3 + \ell_{n-1}^3) V'''_{n-1} + O(\epsilon^5) \ . \tag{2.8}$$

Squaring (2.7) and substituting it into the left-hand side of the above equation, we have

$$(v_n - v_{n-1})(V'_{n-1} + \frac{1}{3} V''_{n-1} \ell_n)\epsilon_n + \frac{1}{4}(\epsilon_n + \epsilon_{n-1})^2 V'^2_{n-1} + O(\epsilon^4) = 0 \ . \tag{2.9}$$

Substituting again (2.7) into (2.9), we derive

$$\epsilon_n - \epsilon_{n-1} = -\frac{2}{3} \frac{V''_{n-1}}{V'_{n-1}} \epsilon_n \ell_n + O(\epsilon^3) \ . \tag{2.10}$$

Hence in the limit as $\epsilon \to 0$, (2.10) becomes

$$\frac{d}{dn}\left[\ln \epsilon_n + \frac{2}{3} \ln V'_n\right] = 0 \ ,$$

or, in terms of (2.4),

$$\left(\frac{dt}{d\tau}\right)^3 \left(\frac{dV}{dx}\right)^2 = \text{constant} \ , \tag{2.11}$$

which describes the asymptotic distribution of $t_n$ versus $n$. In the same limit, either (2.7) or (2.9) gives Newton's equation

$$\frac{d^2x}{dt^2} = -\frac{dV}{dx} \ . \tag{2.12}$$

The constant in (2.11) is determined by the trajectory $x(t)$ and the condition that when $\tau$ varies from 0 to T, t also changes from 0 to T. In the usual continuum mechanics, only Newton's equation is retained. Therefore, the discrete mechanics contains more information than the usual continuum mechanics.

It is of interest to examine the distribution $t(\tau)$ near the point $V'(x) \equiv dV/dx = 0$, since $dt/d\tau$ becomes singular there. Suppose $V'(x) = 0$ at $x = \bar{x}$. Let the particle trajectory in the continuum limit be $x = x(t)$. When $x = \bar{x}$, we have $V'(\bar{x}) = 0$ and, for the solution under consideration, $t = \bar{t}$ so that $\bar{x} = x(\bar{t})$. In the neighborhood $x$ near $\bar{x}$, we may write, with $\dot{x} \equiv dx/dt$,

$$\begin{aligned} V'(x) &\cong (x - \bar{x})\ V''(\bar{x}) \\ &= (t - \bar{t})\ \dot{x}(\bar{t})\ V''(\bar{x}) \quad . \end{aligned}$$

Substituting this expression into (2.11), we find

$$(t - \bar{t}) \propto (\tau - \bar{\tau})^{3/5} \quad .$$

Hence, as $\tau \to \bar{\tau}$, although $dt/d\tau \to \infty$ one sees that $t \to \bar{t}$ and remains finite. Information such as this is lost if one concentrates only on Newton's equation.

## References


1. T. D. Lee, Phys.Lett. 122B, 217 (1982); Shelter Island II, ed. R. Jackiw, N. Khuri, S. Weinberg and E. Witten (Cambridge, the MIT Press, 1985), p. 38-64.
2. "Difference Equations As the Basis of Fundamental Physical Theories", to be published in the proceedings of the symposium Old and New Problems in Fundamental Physics, Pisa, Italy, 1984; "Physics in Terms of Difference Equations", to be published in the proceedings of the Niels Bohr Centenary Symposium, Copenhagen, Denmark, 1985.

# CARTAN DETERMINANTS, LIE ALGEBRA EXTENSIONS, AND THE EXCEPTIONAL GROUP SERIES

Richard H. Capps
Department of Physics
Purdue University
West Lafayette, Indiana 47907

## 1. INTRODUCTION

In this note I utilize the determinant of the generalized Cartan matrix for candidate Dynkin systems for two purposes. The first is to provide an uncomplicated criterion for classifying candidate one-root extensions of diagrams for semisimple Lie algebras. The second is to help determine some important properties of related Lie algebras and their representations.

There are several reasons for believing that some members of the exceptional group series $E_3$, $E_4$,...$E_8$ will be important in future theories of particles. Some regularities in the properties of the members of this series are exhibited in Sec. 4.

## 2. CLASSIFICATION OF CANDIDATE EXTENSIONS

I consider a set of real or hypothetical non-zero root vectors in a Euclidean space. It is required that each pair of vectors $R_j$ and $R_k$ satisfy the dot-product conditions,

$$R_j \cdot R_k = -N\, R_j^2/2 = -M\, R_k^2/2 \quad , \qquad (1)$$

where M and N are non-negative integers. If the vectors are

of different length, I take $R_j^2 \geq R_k^2$, so that $M \geq N$. If M and N are positive the vectors are said to be "connected". Any proposed set of vectors satisfying these conditions, such that the assumed length ratios are consistent, is called a candidate Dynkin system.

Every candidate system and its Dynkin diagram fall into one of three classes, labeled positive, zero, or negative. The system is positive if a set of linearly independent vectors exists that satisfies the requirements. In this case the vectors may be taken as the simple roots for a semisimple Lie algebra.[1] The system is zero if the vectors exist but are linearly dependent. If a system is negative, no set of vectors satisfies all the requirements. The zero diagrams are useful for cataloguing several types of structures, including Kac-Moody algebras[2,3] and replacement bases.[4,5] The only pairs of integers (M, N) that occur in positive systems are (0, 0), (1, 1), (2, 1), and (3, 1). (Thus N is either 0 or 1.) The zero systems include these (M,N) pairs, together with (4,1) and (2,2).

Only one-root extensions of positive Dynkin diagrams are considered here; it can be shown that all indecomposable positive and zero diagrams may be obtained in this way. The original n-root diagram may be decomposable. The new root must satisfy Eq. (1) with each old root; only systems are considered where the final (n + 1)-root diagram is indecomposable. I will give an illuminating proof of an unproved assertion of Ref. (5), that the signature (+, 0, or -) of the extended diagram is the signature of the determinant of the (n + 1)-dimensional Cartan matrix of the diagram. The extending root is normalized to length $\sqrt{2}$; the system may include both longer and shorter roots.

First I present an alternate classification procedure, one with obvious validity. The proposed new root $R_x$ is written as a sum of a piece that is parallel to the n-dimensional subspace of the original simple roots and an orthogonal piece that lies in a new dimension. The dot product conditions of Eq. (1) are used to determine the parallel piece P. Clearly the signature (+, 0, or -) of the extended diagram is the signature of the quantity $2 - P^2$. If $P^2 < 2$, the orthogonal piece may be chosen so that $R_x^2 = 2$, while if $P^2 > 2$ the vector system does not exist.

The length-squared of P is given by the standard formula,[6]

$$P^2 = \Sigma_{ij}\, a_i a_j G_{ij} \ , \tag{2}$$

where G is the symmetric, metric tensor of the algebra, and $a_i$ is the Dynkin component,

$$a_i = (P \cdot R_i)(2/R_i^2) \ . \tag{3}$$

For reasons to be made clear later it is convenient to write $G_{ij}$ in terms of elements of the inverse matrix $V = G^{-1}$, so that the signature factor is

$$2 - P^2 = 2 - \Sigma_{ij} a_i a_j C_{ij}/|V| \ , \tag{4}$$

where $|V|$ is the determinant of V and $C_{ij}$ is the ij cofactor of V. This matrix is related to the Cartan matrix A by

$$V = 2(R^2)^{-1} A \ , \tag{5}$$

where $R^2$ is the diagonal root-length matrix $(R^2)_{ij} = \delta_{ij} R_i^2$, and[6]

$$A_{ij} = (R_i \cdot R_j)(2/R_j^2) \ . \tag{6}$$

It is seen from Eqs. (1) and (3) that the $a_i$ are given by,

$$a_i = -N_i R_\ell^2 / R_i^2 \ , \tag{7}$$

where $R_\ell$ is the longer of the roots $R_i$ and $R_x$, if they are unequal. (The extension prescription of Refs. 4 and 5 is based on the observation that P is the most negative weight of an irreducible representation of the algebra corresponding to the unextended diagram.)

Whether or not the dot-product conditions for the extended system may be satisfied, they may be used in conjunction with Eq. (6) to define an (n + 1)-dimensional Cartan matrix A', and an (n + 1)-dimensional root-length matrix $(R^2)'$. I define the symmetric matrix V' by $V' = 2\,[(R^2)']^{-1} A'$. It follows from Eq. (6) that

$$V'_{ij} = (R_i, R_j)(2/R_i^2)(2/R_j^2) \ . \tag{8}$$

The first n rows and columns of V' are the V of Eq. (5) while the normalization condition and Eq. (8) imply that $V'_{n+1,n+1} = 2$. Thus one may write

$$V' = \left( \begin{array}{c|c} V & \begin{matrix} V'_{1,n+1} \\ \vdots \\ V'_{n,n+1} \end{matrix} \\ \hline V'_{1,n+1} \cdots V'_{n,n+1} & 2 \end{array} \right) \tag{9}$$

If the determinant of V' is expanded in terms of elements of the last row and column, the result is

$$|V'| = 2|V| - \sum_{i,j} V'_{i,n+1} V'_{j,n+1} C_{ij} , \tag{10}$$

where $C_{ij}$ is the ij cofactor of V. Since $R_{n+1}^2 = 2$, $V'_{i,n+1} = (R_i \cdot R_{n+1})(2/R_i^2)$. It follows from Eqs. (1) and (7) that

$$V'_{i,n+1} = a_i \quad .$$

Comparison of Eqs. (4) and (10) shows that $|V'|$ is equal to $(2 - P^2)$ multiplied by the positive determinant $|V|$ associated with the original Lie algebra. Hence the signature of $|V'|$ is the signature of the extended diagram. Since $(R')^2$ is a positive definite matrix, the signature of $|V'|$ is the same as that of the determinant of the generalized Cartan matrix $A' = \frac{1}{2}(R^2)'V'$. This completes the proof.

## 3. THE DETERMINANT OF A FOR THE SIMPLE ALGEBRAS

The matrix A is more fundamental than V, partly because A is invariant to the normalization of the root system. I will list some values of the determinant of A here. The notation $|H|$ is used to denote the determinant of A for the algebra H or candidate diagram H. Almost all positive and zero diagrams may be obtained by adding one root that connects to only one root of a smaller diagram. In such a case a simplied version of the argument leading to Eq. (10) yields the formula.

$$|H_+| = 2|H| - MN|H_-| \quad , \qquad (11)$$

where $H_+$ is the extended system, H the original system and $H_-$ corresponds to the diagram obtained by removing from the original diagram the root that connects to the new root. The values of M and N refer to the new root connection; in all cases considered $N = 1$. If $M \neq 1$, Eq. (11) applies no matter which of the new root and connecting old root is longer.

One may use Eq. (11) to obtain the Cartan determinants

of all simple algebras very quickly. They are

$$|A_n| = n + 1$$

$$|B_n| = |C_n| = 2,$$

$$|D_n| = 4, \quad (n \geqq 2),$$

$$|E_n| = 9 - n \; (3 \leqq n \leqq 8),$$

$$|F_4| = |G_2| = 1 \;,$$

where the semisimple algebras $D_2(A_1 \times A_1)$ and $E_3(A_2 \times A_1)$ are also included.

It is seen from these equations that the only algebras with unit Cartan determinants are $G_2$, $F_4$, and $E_8$ and their direct products. From this fact and Eq. (11) it is seen immediately that these three simple algebras cannot be extended to larger simple algebras.

## 4. THE EXCEPTIONAL GROUP SERIES

The Dynkin diagram for the exceptional group $E_8$ is given below.

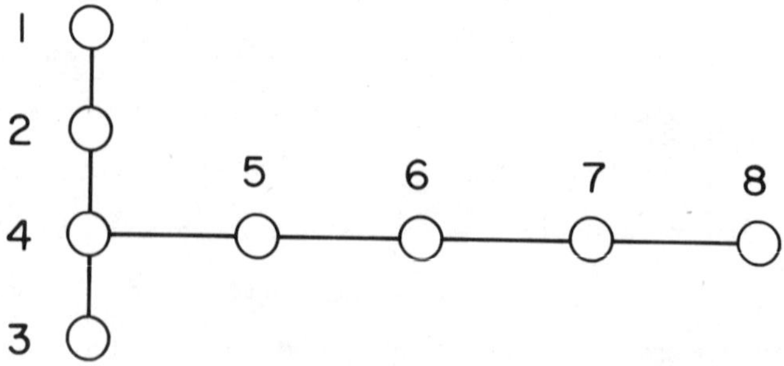


Fig. 1

Successive removals of the simple roots labeled 8 through 4 yields the diagrams for the algebras $E_7$ through $E_3$. The algebras $E_5$, $E_4$, and $E_3$ are identical, respectively, with $D_5$[SO(10], $A_4$[SU(5)], and $(A_2 \times A_1)$ [SU(3) x SU(2)].

There are several reasons for believing that this series may be important in particle physics. For many years people have proposed $E_6$ as a unifying group, using as an argument the fact that $E_5$ and $E_4$ are the attractive unification groups SO(10) and SU(5). Equally cogent is the fact that $E_3$ may correspond to the low-energy group $SU(3)_{color} \times SU(2)_{weak}$. The recent development of the heterotic string theory suggests that the basic symmetry at ultra high energies may be $E_8 \times E_8$.[7]

The importance of the various $E_n$ groups depends on the dynamics and on the form of the symmetry breaking. No specific dynamical model will be considered here. However, it is clear that the concept of the $E_8 \to E_3$ series is not likely to be useful unless there are some regularities that persist throughout the series. I will exhibit a few such regularities here.

The length-squared S of an irreducible representation (irrep.) is defined as the length-squared of the longest weights in the irrep. These include the most positive and most negative weights. The basic irrep. $\Psi_i$ is that with Dynkin indices (components of the most positive weight) $a_j = \delta_{ij}$. All simple roots of $E_n$ may be taken of length $\sqrt{2}$, in which case $G = A^{-1}$. It then follows from the length formula [Eq. (2)] and the inverse formula that the length-squared of the basic irrep. $\Psi_i$ is given by

$$S_i = |H_{-i}|/|H| \quad ,$$

where $H_{-i}$ is the Cartan determinant for the algebra obtained by removing the simple root i from the Dynkin diagram of H. It is easy to see that the values of S for the basic irreps. corresponding to the roots labeled 1, 2, 3, and 4 in Fig. 1 are given by the simple formulas,

$$S_1 = 4/(9-n), \qquad S_2 = 2(n-1)/(9-n),$$

$$S_3 = n/(9-n), \qquad S_4 = 6(n-3)/(9-n),$$

where $3 \leqq n \leqq 8$. Of course the representation $\Psi_4$ does not exist for $E_3$.

In Refs. 4 and 5 a method was given for obtaining a simple root set for $E_n$ (n = 6, 7, or 8) by extending a simple root set of $A_{n-1}$ or $D_{n-1}$. The same procedure is valid for n = 3, 4, and 5; I will demonstrate this here for n = 3. I first describe briefly the extension method. The parallel piece (defined in Sec. 2) of the extending root is the most negative state of the appropriate irrep. (extending irrep) of the original algebra. The perpendicular piece in the new dimension is chosen so that the new root is of length $\sqrt{2}$.

In the $A_{n-1} \rightarrow E_n$ extension the extending irrep. is either the antisymmetric three-quark or antisymmetric three-antiquark irrep. In the case n = 3, this is the singlet irrep. of $A_2$, so the parallel piece is zero. Thus the extending root is orthogonal to the $A_2$ roots, and the new root set is that of $A_2 \times A_1$ $(E_3)$.

In the $D_{n-1} \rightarrow E_n$ extension, the extending irrep. is one of the two spinor irreps. In the case of $D_2$ in the standard basis, the simple roots are (1 -1) and (1 1), which are orthogonal ($D_2 = A_1 \times A_1$). The most negative states of the two spinors are $(-\frac{1}{2}\ \frac{1}{2})$ and $(-\frac{1}{2}\ -\frac{1}{2})$. Each connects with exactly one of the $D_2$ roots, yielding $A_2 \times A_1$.

If the $E_n$ series is important physically, the $E_{n-1} \to E_n$ extensions may be important, where $4 \leqq n \leqq 8$. The extending root for $E_{n-1} \to E_n$ is that labeled n in Fig. 1. In the case $E_3 \to E_4$ the extending irrep. is the direct product of fundamental irreps. of $A_1$ and $A_2$. When $n \geqq 5$ the extending irrep. is the irrep. conjugate to the basic irrep. corresponding to the connecting root of $E_{n-1}$.[4,5] In all the cases the length-squared of the extending irrep. is given by $S = (11 - n)/(10 - n)$.

If an algebra H can be extended to an algebra $H_+$ of one higher rank, a problem that is important for many dynamical models is identifying the irreps. of $H_+$ that contain singlets of H. (For example, this is important if the $H_+$ symmetry breaks down to H symmetry at some energy scale.) Here, two types of singlet states are considered, roots of zero length and "outer states" of an $H_+$ irrep., where the outer states are all those with the maximum weight length.

It is clear that for any such extension there is exactly one zero root of $H_+$ that is a singlet of H, the root corresponding to the direction orthogonal to the simple roots of H. The rule is not so trivial for the outer-state singlets. In the extension scheme of Ref. 4 the shortest H irrep. in the outer states of an $H_+$ irrep. occurs in either the highest or lowest state of the $H_+$ irrep. If one uses this fact and applies Eq. (3) to the highest states of irrep. (rather than to P), it can be shown that the Dynkin indices of the irreps. containing outer-state singlets are all zero except one. The non-zero index corresponds either to the extending root or its conjugate root. I will list two examples, using the root numbering convention of Fig. 1. The irreps. of $E_6$ whose outer states contain $E_5$ singlets are (0 0 0 0 0 k) and (k 0 0 0 0 0), where k is any positive integer. The irreps. of $E_5$ whose outer states contain $E_4$ singlets are (0 0 0 0 k) and (0 0 k 0 0). [For k = 1 and 2 these are the 16 and 126 of SO(10).]

This work was supported in part by the U.S. Department of Energy.

## References


1. Howard Georgi, Lie Algebras in Particle Physics, (Benjamin/ Cummings, Reading, Massachusetts, 1982), p. 177.
2. An early paper listing the zero diagrams is S. Berman, R. Moody, and M. Wonenburger, Indiana Univ. Math. J. 21, 1091 (1972).
3. David Olive, Lectures on Gauge Theories and Lie Algebras, given at the Univ. of Virginia, Fall 1982. See also D. Olive and Neil Turok, Nucl. Phys. B215, 470 (1982).
4. Richard H. Capps, Purdue preprint PURD-TH-85-11.
5. Richard H. Capps, Jour. Math. Phys. (April/May 1986).
6. The standard formulas used in this section are taken from R. Slansky, Phys. Reports 79, 1 (1981), pp. 27 and 28.
7. Gross, Harvey, Martinec, and Rohm, Nucl. Phys. B256, 253 (1985).

# WKB WAVE FUNCTION FOR MANY-VARIABLE SYSTEMS

B. Sakita and R. Tzani
*Department of Physics, City College of the City University of New York, New York, New York 10031, USA*

The WKB method is a non-perturbative semi-classical method in quantum mechanics. The method for a system of one degree of freedom is well known and described in standard textbooks.

The method for a system with many degrees of freedom especially for quantum fields [1] is more involved. There exist two methods; Feynman path integral and Schrödinger wave function. The Feynman path integral WKB method is essentially a stationary phase approximation for Feynman path integrals [1]. The WKB Schrödinger wave function method is on the other hand an extension of the standard WKB to many-variable systems [2].

Several years ago [2] J.-L. Gervais and one of the authors developed a general WKB wave function method for many-variable systems based on an earlier work of Banks, Bender, and Wu [3]. Unfortunately, the derivation of WKB wave function (2.42) in that paper is rather difficult to follow. In this paper we present a simple alternative derivation.

Since the many-variable WKB method is useful for many branches in physics to which Professor Gyo Takeda's interests extend and since Professor Takeda loves to discuss physics using Schrödinger wave function, we believe it is appropriate to present this paper on the occasion of his sixtieth birthday.

We consider a system with N degrees of freedom. The Lagrangian is given by

$$L=\frac{1}{2}\dot{\vec{q}}^{2}-V(\vec{q}) \tag{1}$$

where, $q^i$ are the coordinates and $V(\vec{q})$is a potential. We assume V depends on a parameter g (coupling constant) as following:

$$V(\vec{q})=\frac{1}{g^2}U(g\vec{q}) \tag{2}$$

Let $\vec{q}_0(t)$ be a solution of the classical equation

$$\ddot{q}^i+\frac{\partial V(\vec{q})}{\partial q^i}=0 \tag{3}$$

The form (2) assures that $\vec{q}_0(t)$ is of the order $\frac{1}{g}$.

The classical solution $\vec{q}_0(t)$ specifies a line in the configuration space, which we call the classical trajectory. The essence of many-variable WKB method is to consider the Schrödinger wave function in the vicinity of the classical trajectory. This was done in G.S. with a change of variables:

$$q^i \quad (i=1,2,...N) \rightarrow t\,,\ \eta^a \quad (a=1,...N-1)$$

such that

$$q^i=q^i_0(t)+n^i_a(t)\eta^a \tag{4}$$

where $n^i_a(t)$ is a set of N-1 independent orthonormal vectors normal to the tangent of the classical trajectories.

In this paper we choose a set of slightly different variables; $\tau$, $\xi^m\,(m=1,2,...N-1)$

$$q^i=q^i_0(\tau)+v^i_m(\tau)\xi^m \tag{5}$$

where $\vec{v}_m(t)$ are N-1 independent solutions of the following small fluctuation equation:

$$(\,\partial_t^2\delta_{ij}+U_{ij}\,)\;v^j_m(t)=0 \tag{6}$$

$$U_{ij}(t)\equiv\frac{\partial^2V(\vec{q})}{\partial q^i\partial q^j} \qquad \textit{evaluated at} \quad \vec{q}=\vec{q}_0(t) \tag{7}$$

We note $q_0(\tau)\approx O(\frac{1}{g})$ and $v^i_m(\tau)\approx O(1)$. As we shall see, the advantage of the new choice of variables is due to the fact that equation (5) consists of a set of classical trajectories, namely (5) for a given $\xi$ is an approximate solution of the classical equation (3).

We begin with the Schrödinger equation

$$\left[-\frac{1}{2}\frac{\partial^2}{\partial q^{i2}}+V(\vec{q})\right]\Psi(\vec{q}) = E\Psi(\vec{q}) \tag{8}$$

As usual we set

$$\Psi(\vec{q}) = e^{iS(\vec{q})} \tag{9}$$

We then split S as

$$S=S_0+S_1 \tag{10}$$

where $S_0$ is a solution of the following Hamilton-Jacobi equation

$$\frac{1}{2}\left[\frac{\partial S_0}{\partial q^i}\right]^2+V(\vec{q}) - E_0 = 0 \tag{11}$$

Then the Schrödinger equation becomes

$$\frac{1}{2}\left[\frac{\partial S_1}{\partial q^i}\right]^2 - \frac{i}{2}\frac{\partial^2 S_1}{\partial q^{i2}} + \frac{\partial S_0}{\partial q^i}\frac{\partial S_1}{\partial q^i} - \frac{i}{2}\frac{\partial^2 S_0}{\partial q^{i2}} - (E-E_0) = 0 \tag{12}$$

We solve equation (11) to obtain $S_0$ and then solve equation (12) to obtain $S_1$.

Using (5) we obtain

$$V(\vec{q})=V(\vec{q}_0) + \frac{\partial V}{\partial q_0^i}\xi^i + \frac{1}{2}U_{ij}\xi^i\xi^j + O(g) =$$

$$= V(\vec{q}_0) - \dot{\vec{q}}_0\dot{\vec{\xi}} - \frac{1}{2}\ddot{\vec{\xi}}\vec{\xi} + O(g) \tag{13}$$

where we used the notation

$$\xi^i(\tau) \equiv v^i_m(\tau)\xi^m \tag{14}$$

and

$$\ddot{\vec{q}}_0(\tau)\,\vec{\xi} = \dot{\vec{q}}_0\,\dot{\vec{\xi}} \tag{15}$$

which can be proven by considering the Wronskian associated with the differential equation (6).

We insert (13) into the Hamilton-Jacobi equation (11) and choose $E_0$ to be the energy associated with the classical solution $\vec{q}_0(t)$

$$\frac{1}{2}\dot{\vec{q}}_0^{\,2} + V(\vec{q}_0) = E_0 \tag{16}$$

Then we obtain

$$\frac{1}{2}\left[\frac{\partial S_0}{\partial q^i}\right] - \frac{1}{2}\dot{\vec{q}}_0^{\,2} - \dot{\vec{q}}_0\,\dot{\vec{\xi}} - \frac{1}{2}\dot{\vec{\xi}}\,\dot{\vec{\xi}} + O(g) = 0 \tag{17}$$

This is obviously satisfied if we set

$$\frac{\partial S_0}{\partial q^i} = \dot{q}_0^i(\tau) + \dot{\xi}^i - \frac{\dot{q}_0^i}{2\dot{\vec{q}}_0^{\,2}}(\dot{\vec{\xi}}^2 - \ddot{\vec{\xi}}\vec{\xi}) + O(g^2) \tag{18}$$

The first, the second and the third terms are of order $\frac{1}{g}$, 1, and g respectively. Multiplying by $\dot{q}^i \equiv \frac{\partial q^i}{\partial \tau} = \dot{q}_0^i + \dot{\xi}^i$ we obtain

$$\frac{\partial S_0}{\partial \tau} = \dot{\vec{q}}_0^{\,2} + \frac{\partial}{\partial \tau}\left(\dot{\vec{q}}_0\vec{\xi} + \frac{1}{2}\vec{\xi}\,\dot{\vec{\xi}}\right) \tag{19}$$

where we used (15). So, by integrating we obtain

$$S_0 = \int^{\tau} \dot{\vec{q}}_0^{\,2}(t)\,dt + \dot{\vec{q}}_0(\tau)\,\vec{\xi}(\tau) + \frac{1}{2}\,\vec{\xi}(\tau)\,\dot{\vec{\xi}}(\tau) \tag{20}$$

In this integration we obtain an additional $\xi$ dependent term (integratior constant), but it can be shown to be zero by constructing $\frac{\partial S_0}{\partial \xi^m}$ if w multiply (18) by $v_m^i \equiv \frac{\partial q^i}{\partial \xi^m}$.

Next we solve (12) for $S_1$. For this purpose we first express $\frac{\partial}{\partial q^i}$ in terms of $\frac{\partial}{\partial \tau}$ and $\frac{\partial}{\partial \xi^m}$. Let us define

$$v_0^i(\tau) \equiv \dot{q}_0^i(\tau) \tag{21}$$

and extend the index m to include $m = 0$. Note $v_0^i \approx O(\frac{1}{g})$ and $v_m^i$ $(m \neq 0) \approx O(1)$. Define the inverse matrix $v^{-1}{}_i^m$ by

$$v^{-1}{}_i^m\, v_m^j \equiv \delta_{ij} \tag{22}$$

Note $v^{-1}{}_i^0 \approx O(g)$ and $v^{-1}{}_i^m (m \neq 0) \approx O(1)$. To order g we obtain

$$\frac{\partial}{\partial q^i} = \nu^{-1}{}_i^0 (1 - \dot{\xi}^i \nu^{-1}{}_i^0) \frac{\partial}{\partial \tau} + (\nu^{-1}{}_i^m - \nu^{-1}{}_i^0 \dot{\xi}^i \nu^{-1}{}_i^m) \frac{\partial}{\partial \xi^m} \tag{23}$$

An important point to notice in (23) is that the coefficient of $\frac{\partial}{\partial \tau}$ is at most of order g. So, if we keep the terms up to O(1) in (12) the first two terms in (12) contain only the first and the second derivatives of $S_1$ with respect to $\xi^m$: $\frac{\partial S_1}{\partial \xi^m}$, $\frac{\partial^2 S_1}{\partial \xi^m \partial \xi^m}$.

Next we evaluate the third and the fourth terms. Using (18) we obtain

$$\frac{\partial S_0}{\partial q^i} \frac{\partial S_1}{\partial q^i} \approx \frac{\partial q_i}{\partial \tau} \frac{\partial S_1}{\partial q^i} = \frac{\partial S_1}{\partial \tau} \tag{24}$$

and

$$\frac{\partial^2 S_0}{\partial q^{i\,2}} = \frac{\partial}{\partial q^i} \left( \frac{\partial S_0}{\partial q^i} \right) \approx \frac{\partial}{\partial q^i} ( \dot{q}_0^i + \dot{\eta}^i )$$

$$\approx \nu^{-1}{}_i^0 \ddot{q}_0^i + \nu^{-1}{}_i^m \dot{v}_m^i = \nu^{-1}{}_i^0 \dot{v}_0^i + \nu^{-1}{}_i^m \dot{v}_m^i$$

$$= \frac{\partial}{\partial \tau} \ln \det \nu \tag{25}$$

which is independent of $\xi$.

Thus, an obvious solution for $S_1$ is given by

$$\frac{\partial S_1}{\partial \xi^m} = 0 \tag{26}$$

and

$$\frac{\partial S_1}{\partial \tau} - \frac{i}{2} \frac{\partial}{\partial \tau} \ln \det \nu - (E - E_0) = 0 \tag{27}$$

Accordingly,

$$S_1 = (E - E_0)\tau + \frac{i}{2} \ln \det \nu \tag{28}$$

If we combine (9), (10), (20), and (28) together we obtain the following expression for the WKB wave function

$$\Psi = \frac{1}{(\det v)^{1/2}} \exp i \left| \int^{\tau} \dot{\vec{q}}_0^{\,2}(t)dt + E_1\tau + \dot{\vec{q}}_0(\tau)\vec{\xi}(\tau) + \frac{1}{2}\vec{\xi}\,\dot{\vec{\xi}} \right| \quad (29)$$

where

$$E_1 = E - E_0 \quad (30)$$

**Appendix**

In this Appendix we show that our WKB wave function (29) is really the same as (2.42) of G.S.

The relation between the variables of the two methods is given by (4) and (5):

$$\vec{q}_0(\tau) + \vec{v}_m(\tau)\xi^m = \vec{q}_0(t) + \vec{n}_a(t)\eta^a \quad (31)$$

where

$$\vec{n}_a\vec{n}_b = \delta_{ab}, \qquad \vec{n}_a\dot{\vec{q}}_0 = 0 \quad (32)$$

Let

$$\tau = t - \delta t, \qquad \xi^m = w_a^m\eta^a - \delta\eta^m \quad (33)$$

where $\delta t$ and $\delta\eta \approx O(g)$. Inserting (33) into the left hand side of (31) and expanding $\vec{q}_0(\tau)$ and then multiplying both sides by $\dot{\vec{q}}_0(t)$ we obtain

$$-\dot{\vec{q}}_0^{\,2}(t)\delta t + \frac{1}{2}\dot{\vec{q}}_0 \cdot \ddot{\vec{q}}_0(\delta t)^2 + \dot{\vec{q}}_0\vec{\eta} - \dot{\vec{q}}_0\dot{\vec{\eta}}\delta t - \dot{\vec{q}}_0\delta\vec{\eta} = 0 \quad (34)$$

where we used the notation

$$\vec{\eta} \equiv \vec{v}_m w_a^m\eta^a, \qquad \dot{\vec{\eta}} \equiv \dot{\vec{v}}_m w_a^m\eta^a \quad (35)$$

From (34) we obtain

$$\delta t \approx \frac{\dot{\vec{q}}_0\vec{\eta}}{\dot{\vec{q}}_0^{\,2}} \quad (36)$$

Next we expand $\int^{\tau} \dot{\vec{q}}_0^{\,2}(t)dt$ and $\dot{\vec{q}}(\tau)\vec{\xi}(\tau)$ and use (34) and (35) to obtain

$$\int^{\tau} \dot{\vec{q}}_0^{\,2}(t)dt + \dot{\vec{q}}(\tau)\vec{\xi}(\tau) \approx \int^{t} \dot{\vec{q}}^{\,2}(\tau)dt + \frac{1}{2}\dot{\vec{q}}_0\ddot{\vec{q}}_0(\delta t)^2 - \dot{\vec{q}}_0\dot{\vec{\eta}}\delta t \quad (37)$$

$$\int^{T} \dot{\vec{q}}_0^{\,2}(t)dt + \dot{\vec{q}}(\tau)\vec{\xi}(\tau) \approx \int^{t} \dot{\vec{q}}^{\,2}(\tau)dt + \frac{1}{2}\dot{\vec{q}}_0\ddot{\vec{q}}_0(\delta t)^2 - \ddot{\vec{q}}_0\vec{\eta}\delta t \qquad (37)$$

We now evaluate

$$\frac{1}{2}\vec{\xi}(\tau)\dot{\vec{\xi}}(\tau) \approx \frac{1}{2}\vec{\eta}(t)\dot{\vec{\eta}}(t) = \frac{1}{2}\frac{(\dot{\vec{q}}_0\vec{\eta})(\dot{\vec{q}}_0\dot{\vec{\eta}})}{\dot{\vec{q}}_0^{\,2}} + \frac{1}{2}(\vec{\eta}\vec{n}_a)(\dot{\vec{\eta}}\vec{n}_a) \qquad (38)$$

where we used the completeness relation of $\vec{n}_a$ and $\dot{\vec{q}}_0$. We define matrix $u_m^a$ by expanding $v_m^i(t)$:

$$v_m^i = \frac{\dot{\vec{q}}_0\vec{v}_m}{\dot{\vec{q}}^{\,2}}\dot{q}_0^i + u_m^a n_a^i \qquad (39)$$

It is in general dependent on t. We choose w in (33) and (35) as the inverse of u:

$$u_m^a w_b^m = \delta_{ab} \qquad (40)$$

Then

$$\vec{\eta}\vec{n}_a = \eta^a \qquad (41)$$

$$\dot{\vec{\eta}}\vec{n}_a = \vec{n}_a\dot{\vec{v}}_m w_b^m \eta^b = \left( \frac{\dot{\vec{q}}\dot{\vec{v}}_m}{\dot{\vec{q}}_0^{\,2}}(\ddot{\vec{q}}_0\vec{n}_a) + (Du_m)^a \right) w_b^m \eta^b \qquad (42)$$

where

$$(Du_m)^a = (\partial_t\delta_{ab} + \Gamma_b^a)u_m^b \qquad (43)$$

and

$$\Gamma_b^a = \vec{n}_a\dot{\vec{n}}_b \qquad (44)$$

We put all the above together to obtain

$$\frac{1}{2}\vec{\xi}\dot{\vec{\xi}} \approx$$

$$\approx \frac{1}{2}\frac{(\dot{\vec{q}}_0\vec{\eta})(\dot{\vec{q}}_0\dot{\vec{\eta}})}{\dot{\vec{q}}_0^{\,2}} + \frac{1}{2}\frac{(\dot{\vec{q}}_0\vec{\eta})(\ddot{\vec{q}}_0\vec{\eta})}{\dot{\vec{q}}^{\,2}} -$$

$$-\frac{1}{2}\left|\frac{\dot{\vec{q}}_0\vec{\eta}}{\dot{\vec{q}}_0^{\,2}}\right|^2(\dot{\vec{q}}_0\ddot{\vec{q}}_0) + \frac{1}{2}w_a^m(Du_m)^b\eta^a\eta^b$$

$$= \delta t \dot{\vec{q}}_0 \dot{\vec{\eta}} - \frac{1}{2}(\delta t)^2 \dot{\vec{q}}_0 \ddot{\vec{q}}_0 + \frac{1}{2}\Omega_{ab}\eta^a\eta^b \tag{45}$$

where

$$\Omega_{ab} = w_a^m (Du_m)^b \tag{46}$$

and we used

$$\dot{\vec{q}}_0 \dot{\vec{\eta}} = \ddot{\vec{q}}_0 \vec{\eta} \tag{47}$$

By using (39) it is easy to show

$$\det v = (\dot{\vec{q}}_0^{\,2})^{\frac{1}{2}} \det u + O(g) \tag{48}$$

Inserting (37), (45), and (48) into (29) we obtain

$$\Psi = \frac{1}{(\dot{\vec{q}}^{\,2})^{\frac{1}{4}} (\det u)^{\frac{1}{2}}} \exp i \left[ \int^t \dot{\vec{q}}_0^{\,2} dt + \frac{1}{2}\Omega_{ab}\eta^a\eta^b \right] \tag{49}$$

up to O(1). (49) is the WKB wave function of G.S.

This work is supported in part by NSF grant PHY-82-16364 and PSC-BHE Faculty Research Award RF 6-64266.

**References**

1 *R.Dashen,B.Hasslacher,and A.Neveu,Phys.Rev.* **D10 (1974), 4114**

2 *J.-L.Gervais and B.Sakita,Phys.Rev.* **D16** (1977), 3507 We cite this paper as G.S.

3 *T.Banks,C.M.Bender and T.T.Wu,Phys.Rev.* **D8** (1973), 3346 *T.Banks and C.M.Bender,ibid.* **8** (1973), 3366

# Integral Equation for Coulomb Problem

Tatuya Sasakawa
Department of Physics, Tohoku University
Sendai 980, Japan

For short range potentials an inhomogeneous (homogeneous) Lippmann-Schwinger integral equation of the Fredholm type yields the wave function of scattering (bound) state. For the Coulomb potential, this statement is no more valid. It has been felt difficult to express the Coulomb wave function in a form of an integral equation with the Coulomb potential as the perturbation. In the present paper, we show that an inhomogeneous integral equation of a Volterra type with the Coulomb potential as the perturbation can be constructed both for the scattering and the bound states. The equation yielding the binding energy is given in an integral form. The present treatment is easily extended to the coupled Coulomb problems.

## I. INTRODUCTION

For a short range potential, the wave function of scattering (bound) state is expressed in a form of inhomogeneous (homogeneous) Lippmann-Schwinger equation of the Fredholm type, with the potential as the perturbation. This statement is not applicable to the Coulomb potential which is characterized by a logarithmic divergence. To overcome this difficulty, West[1] proposed an integral equation with a generalized kernel applicable also to the Coulomb potential. He demonstrated that the first order iteration agrees almost with the exact wave function. However, the kernel of this equation is complicated and we will not be able to perform the perturbation iterations up to any desired order. Reflecting this situation, no text book of quantum mechanics describes the Coulomb problem in a form of an integral equation.

In the present paper, we demonstrate that the scattering and the bound states of a pure Coulomb system can be expressed as an inhomogeneous integral equation of a Volterra type with the Coulomb potential as the perturbation. This integral equation can be solved analytically, and yields the correct Coulomb wave function. For short range potentials, the spherical Bessel and Hankel functions are employed to construct the Green function. This time, we utilize the regular and irregular confluent hypergeometric functions for this purpose.

This work was motivated by the following reasons. Firstly, if the equations are coupled in a more-than-two body Coulomb problem, the usual way of obtaining the binding energy by cutting the wave function at some power of distance should not work, since there are as many wave functions as the number of coupled equations, whereas only one condition should be imposed for giving a binding energy of the system. In the present paper, the author aims at giving the equation for binding energy in an integral form. In this way, in the case of coupled Coulomb equations too, we can obtain a set of coupled integrals that yield the binding energy. Secondly, the use of the family of Bessel function causes divergence or discontinuity, whereas the use of the family of confluent hypergeometric functions does not. Therefore, it is tempting to try a formulation in terms of this family of functions.

Although the aim of the present paper is to serve as a first step of handling coupled Coulomb equations, we restrict ourselves to an uncoupled equation in the present paer. This is for clarity of the content. We present a treatment which can be readily extended to coupled Coulomb problems.

## II. INTEGRAL EQUATION

With obvious notations, we begin with the Schrödinger equation for the Coulomb potential

$$[\frac{d^2}{dr^2} + \frac{2}{r}\frac{d}{dr} - \frac{\ell(\ell+1)}{r^2} + k^2 - \frac{2\eta k}{r}]\psi_\ell(kr) = 0, \qquad (1)$$

where

$$\eta = Z_1 Z_2 e^2/\hbar v \quad . \qquad (2)$$

As usual, we put

$$\psi_\ell(kr) = e^{ikr}(kr)^\ell f_\ell(-2ikr) \qquad (3)$$

and

$$kr = -\frac{z}{2i} \quad . \qquad (4)$$

Then we get the confluent hypergeometric equation

$$[\, z\frac{d^2}{dz^2} + (\, c - z\,)\frac{d}{dz} - a\,]\, f_\ell(z) = i\eta f_\ell(z), \qquad (5)$$

where

$$a = \ell + 1, \qquad c = 2(\,\ell + 1).$$

The solution of Eq.(5) is a confluent hypergeometric function,

$$f_\ell(z) = \Phi(a + i\eta, c; z). \tag{6}$$

In this section, we show that this function is obtained as a solution of an integral equation, with the Coulomb interaction $i\eta/z$ as the perturbation.

Usually, the regular and irregular spherical Bessel functions are used for making the Green function. Here, we employ the confluent hypergeometric functions

$$\Phi(a,c;z) = \frac{\Gamma(c)}{\Gamma(a)} \sum_{n=0}^{\infty} \frac{\Gamma(n + a)}{\Gamma(n + c)} \frac{z^n}{n!} \tag{7a}$$

and

$$\overline{\Phi}(a,c;z) = z^{1-c} \Phi(a - c + 1, 2 - c ; z), \tag{7b}$$

that satisfy the equation without Coulomb interaction,

$$\left[ z \frac{d^2}{dz^2} + ( c - z ) \frac{d}{dz} - a \right] \Phi = 0. \tag{8}$$

Then, we construct an inhomogeneous integral equation for $f_\ell(z)$ of Eq.(5), which is regular at the origin.

$$f_\ell(z) = \Phi(a,c;z) - i\eta\left[ \Phi(a,c;z) \int_0^z \frac{\overline{\Phi}(a,c;z')}{z'W} - \overline{\Phi}(a,c;z) \int_0^z \frac{\Phi(a,c;z')}{z'W} \right] f_\ell(z')dz' \quad . \tag{9}$$

Here W denots the Wronskian

$$W = W(\Phi,\overline{\Phi}) = - ( c - 1 )z^{-c}e^z \quad . \tag{10}$$

Let us define a function $\theta_\sigma(a,c;z)$ by[2]

$$\theta_\sigma(a,c;z) = z^\sigma \sum_{m=0}^{\infty} \frac{\Gamma(\sigma + a + m)\Gamma(\sigma)\Gamma(\sigma + c - 1)}{\Gamma(\sigma + a)\Gamma(\sigma + m + 1)\Gamma(\sigma + c + m)} z^m . \qquad (11)$$

The series on the right hand side is convergent for all values of z.

By making use of Eqs.(4.210) and (4.211) of ref.2, we can show that

$$[ - \Phi(a,c;z) \int_0^z \frac{\bar{\Phi}(a,c;z')}{z'W} + \bar{\Phi}(a,c;z) \int_0^z \frac{\Phi(a,c;z')}{z'W} ] z'^n dz'$$

$$= \theta_{n+1}(a,c;z). \qquad (12)$$

This is a key equation for handling the perturbation iterations of Eq.(9). If we use Eqs.(7a) and (12), we obtain the solution (6), for example, in the following manner. We expand $f_\ell(z)$ in the power series of z;

$$f_\ell(z) = \sum_{n=0}^{\infty} \alpha_n z^n .$$

Then, Eq.(9) reads

$$\sum_{n=0}^{\infty} \alpha_n z^n = \Phi(a,c;z) + i\eta \sum_{n=0}^{\infty} \alpha_n \theta_{n+1}(a,c;z) . \qquad (13)$$

Comparing both sides of this equation, we obtain

$$\alpha_0 = 1 \quad , \alpha_1 = \frac{a + i\eta}{c} , \quad \alpha_2 = \frac{1}{2c(c + 1)}(a + i\eta)(a + 1 + i\eta), \text{etc},$$

and, as a result, we obtain Eq.(6)

$$f_\ell(z) = 1 + \frac{a + i\eta}{c} z + \frac{(a + i\eta)(a + 1 + i\eta)}{2c(c + 1)} z^2 + \ldots$$

$$= \Phi(a + i\eta, c ; z) = {}_1F_1(a + i\eta, c ; z) , \qquad (14)$$

as required.

## III. EQUATION FOR BINDING ENERGY

One method of finding the equation that determines the binding energy is the following. For a large n,

$$\alpha_n z^n = \frac{(a + i\eta)(a + 1 + i\eta)\dots(a + n - 1 + i\eta)}{c(c + 1)\dots(c + n - 1)n!} z^n$$

$$= \frac{(c - 1)!}{(a - 1 + i\eta)!} \frac{(a + n - 1 + i\eta)!}{(c + n - 1)!} \frac{z^n}{n!}$$

$$\sim \frac{\Gamma(c)}{\Gamma(a + i\eta)} \frac{n^{a - c + i\eta}}{n!} z^n$$

$$\sim \frac{\Gamma(c)}{\Gamma(a + i\eta)} \frac{1}{(n + c - a - i\eta)!} z^n$$

$$= \frac{\Gamma(c)}{\Gamma(a + i\eta)} z^{a + i\eta - c} \frac{z^m}{m!}, \qquad (15)$$

where $m = n + c - a + i\eta$. Threrefore,

$$f_\ell(z) = \sum_n \alpha_n z^n \sim \frac{\Gamma(c)}{\Gamma(a + i\eta)} z^{a + i\eta - c} e^z \qquad (16)$$

for a large z. If we put this expression in Eq.(3), we get the asymptotic behavior of $\psi_\ell(kr)$ as

$$\psi_\ell(kr) \sim \frac{\Gamma(2(\ell + 1))}{\Gamma(\ell + 1 + i\eta)} (-2i)^{-(\ell + 1) + i\eta} (kr)^{-1 + i\eta} e^{-ikr}. \qquad (17)$$

When k is on the upperhalf of the imaginary axis, Eq.(17) reads,

$$\psi_\ell(kr) \sim \frac{\Gamma(2(\ell + 1))}{\Gamma(\ell + 1 + i\eta)} (-2i)^{-\ell} (2|k|r)^{-1 + i\eta} e^{|k|r}. \qquad (18)$$

This function exponentially diverges for a large value of r. However, this divergence is prevented if the argument of $\Gamma(\ell + 1 + i\eta)$ is equal to a negative integer - n, in which case $(\Gamma(-n))^{-1}$ vanishes. If we use Eq. (2), this condition amounts to

$$k = -i \frac{1}{(\ell + n + 1)} \frac{Z_1 Z_2 e^2 \mu}{\hbar^2}, \tag{19}$$

where $\mu$ is the reduced mass. Note that for a negative charge $Z_i < 0$. The binding energy of the system is given by

$$|E| = \frac{\hbar^2}{2\mu} |k|^2 = \frac{\mu}{2(\ell + n + 1)^2} \left(\frac{Z_1 Z_2 e^2}{\hbar}\right)^2 . \tag{20}$$

This is a well known formula found in any text book of quantum mechanics.

This method of obtaining the binding energy from a series expansion in powers of distance works only for a system in which the original equation is not coupled as Eq.(1). However, in Nature, there must be a system that should be described by a coupled set of Coulomb equations. For instance, if the original equations consist of two coupled equations, we have two two series of the kind of Eq.(14), but have only one condition to determine the binding energy. This means that by the method which lead to Eq. (20) can not be used in the case of coupled equations. The way out of this difficulty is to have a set of equations for yielding the binding energy expressed in an integral form, or in other words, in an algebraic form not involving distance. In what follows, we show it for (uncoupled) equation (1), or equivalently for Eq.(9). The extension to coupled equations is then straightforward.

For this purpose, we use a function $\Psi(a,c;z)$ defined by[3]

$$\Psi(a,c;z) = \frac{\Gamma(1-c)}{\Gamma(a-c+1)} \Phi(a,c;z) + \frac{\Gamma(c-1)}{\Gamma(a)} \bar{\Phi}(a,c;z) . \tag{21}$$

We put this function in Eq.(9) to obtain

$$f_\ell(z) = \Phi(a,c;z) - i\eta \frac{\Gamma(a)}{\Gamma(c-1)}[\ \Phi(a,c;z) \int_0^z \Psi(a,c;z')$$

$$- \Psi(a,c;z) \int_0^z \Phi(a,c;z')\ ]\ \frac{f_\ell(z')}{z'W}\, dz' \ . \qquad (22)$$

The asymptotic form of $\Phi(a,c;z)$ and $\Psi(a,c;z)$ are known to be[4]

$$\Phi(a,c;z) \sim \frac{\Gamma(c)}{\Gamma(a)}\, e^z z^{a-c} \quad ,$$

$$(23)$$

$$\Psi(a,c;z) \sim z^{-a} \ .$$

Therefore, the asymptotic form of Eq.(3) reads

$$\psi_\ell(kr) \sim (-z/2i)^\ell \{e^{z/2} z^{a-c}[\ 1 - i\eta \frac{\Gamma(a)}{\Gamma(c-1)} \int_0^\infty \Psi(a,c;z')\ \frac{f_\ell(z')}{z'W} dz']$$

$$+ e^{-z/2} z^{-a} (i\eta \frac{\Gamma(a)}{\Gamma(c-1)} \int_0^\infty \Phi(a,c;z')\ \frac{f_\ell(z')}{z'W}\, dz'\} \ .$$

$$(24)$$

For a bound state, $z = 2|k|r$. Therefore, the binding energy is obtained from the equation

$$1 - i\eta \frac{\Gamma(a)}{\Gamma(c-1)} \int_0^\infty \Psi(a,c;z)\ \frac{f_\ell(z')}{z'W}\, dz' = 0 \quad , \qquad (25)$$

where $k = i|k|$. Equation (25) is whant we intended to obtain in the present paper.

Now, let us check if Eq.(25) actually yields the same formula (19) for the binding energy. The evaluation of the integral in Eq.(25) is done in the following manner.

$$\int_0^\infty \Psi(a,c;z)\,\frac{f_\ell(z)}{zW}\,dz$$

$$= -\frac{1}{c-1}\int_0^\infty e^{-z}z^{c-1}\,\Psi(a,c;z)\,\Phi(a+i\eta,c;z)\,dz$$

$$= -\frac{1}{c-1}\sum_{n=0}\alpha_n\int_0^\infty e^{-z}\,z^{n+c-1}\,\Psi(a,c;z)\,dz$$

$$= -\frac{1}{c-1}\sum_{n=0}\alpha_n\,\frac{\Gamma(1+n)\Gamma(n+c)}{\Gamma(1+a+n)}\quad. \tag{26}$$

At the last stage, we have used the formula[5]

$$\int_0^\infty e^{-ax}x^{s-1}\,(b,d;\mu x)dx = \frac{\Gamma(1+s-d)\Gamma(s)}{\Gamma(1+b+s-d)}\,a^{-s}F(b,s;1+b+s-d;1-\mu/a),$$

$$\mathrm{Re}\ s > 0,\ 1+\mathrm{Re}\ s > \mathrm{Re}\ d. \tag{27}$$

Finally, the left side of Eq.(25) becomes

$$1+i\eta\,\frac{\Gamma(\ell+1)}{\Gamma(2\ell+1)}\,\frac{1}{(2\ell+1)}\sum_{n=0}\alpha_n\,\frac{\Gamma(1+n)\Gamma(2(\ell+1)+n)}{\Gamma(\ell+2+n)}$$

$$= 1+i\eta\,\frac{\Gamma(\ell+1)}{\Gamma(2(\ell+1))}\,\{\alpha_o\,\frac{\Gamma(1)\Gamma(2\ell+2)}{\Gamma(\ell+2)} + \alpha_1\frac{\Gamma(2)\Gamma(2\ell+3)}{\Gamma(\ell+3)} + \ldots$$

$$= 1+\frac{i\eta}{\ell+1}+\frac{i\eta(\ell+1+i\hbar)}{(\ell+1)(\ell+2)}+\frac{i\eta(\ell+1+i\eta)(\ell+2+i\eta)}{(\ell+1)(\ell+2)(\ell+3)}+\ldots$$

$$= \prod_{n=1}^{\infty}\left(1+\frac{i\eta}{\ell+n}\right)\quad. \tag{28}$$

Setting this expression equal to zero, we obtain Eq.(19) for the binding energy.

Finally, we mention that the Coulomb wave function of the scattering state[6] is also given by Eq.(3) with Eq.(6), the solution of Eq.(9). [As a convention, $\Phi(a-i\eta,\ c;\ 2ikr)$ is used for the scattering state.]

## IV. CONCLUSION

We have obtained the integral equation in which a Coulomb potential is treated as the perturbation. The bound state as well as the scattering state satisfy the inhomogeneous equation of the Volterra type. Also, we obtained the equation for yielding the binding energy in an integral form. The extension to coupled Coulomb equations is easily done. By this work, the calculation of the binding energy of a coupled Coulomb problem becomes now feasible.

A collaboration with Dr. B. Talukdar at an early stage is acknowledged.

References

1) G.B. West, J. Math. Phys. 8, 942 (1967).

2) A.W. Babister, Transcendental Functions Satisfying Nonhomogeneous Linear Differenetial Equations (McMillan, New York, 1967).

3) A. Erdélyi et al., Higher Transcendental Functions (McGraw Hill, 1953), p. 257.

4) A. Erdélyi et al., ibid. p. 278.

5) A. Erdélyi et al., ibid. p. 287.

6) For example, M. Abramowitz and I.A. Stegun, Handbook of Mathematical Functions (Dover, 1964), p. 537.

# Number of Possible Algorithms

**Tatsuo Tsukamoto**

Department of Physics, Tohoku University,
Aramaki aza Aoba, 980 Sendai, Japan

**Hisao Matsuzaki**

Department of Physics, Tohoku College of Pharmacy,
Komatsushima, 983 Sendai, Japan

**Yoshiyuki Kawazoe**

Education Center for Information Processing, Tohoku University,
Kawauchi, 980 Sendai, Japan

Abstract:

The number of binary trees for a fixed number of leaves is derived in a closed form. The result gives the number of possible algorithms of executing an expression composed of binomial operators. This is also the number of possible immediate constituent structures for a sentence with fixed length.

## 1. Introduction

This paper is dedicated to the 60th anniversary of Professor Gyo Takeda, who always criticizes that nowadays everybody depends too much on computers. However, we hope that his wide interest permits us to contribute the present work on computer science. Our concern is not limited to numerical calculations. Searches for the best strategy in computer programming often encounter the old and familiar question, "What is intelligence?" Although our study starts from algorithms, it may cover wider range of interest.[1)] For example, it is closely related to the syntax structure of natural languages.

One of the main aims of computer science is evidently to improve the computing speed. To achieve this, programmable logic unit (PLU) has been introduced to realize a hardware mapping of the inverse Polish notation. [2)] To test the efficiency of mapping program to PLU, it is necessary to count explicitly the number of binary trees, which directly expresses the possible combinations of variables with binomial operators. Present paper will partly solve this problem.

## 2. Possible Algorithms

To compute an expression composed of more than three terms, one has to choose a specified algorithm out of many possible ones. Actually evaluation of a+b+c will be performed in two ways; (a+b)+c or a+(b+c), under the assumption that the processes should flow from left to right. These two are absolutely different in the stand point of algorithmical procedures. In the former case we need only one stack, and in the latter we need two. Good optimization will meet the problem of competing two requirements. It must shorten the time of computation within the limitation of a given number of parallel processors. At the same time, it must take account of the number of available stacks. Even if the technological development is so high, it is always desirable to reduce the necessary amount of hardware. For example, among the five possible procedures with four terms, which are shown below as binary trees, the one which needs the least number of stacks is ab+c+d+. We use expressions written in the inverse Polish notation, where ab+ means a+b, and ab+c+d corresponds to the traditional expression with parentheses ( (a+b)+c)+d, which actually needs only one stack. The quickest procedure is ab+cd++, i.e. (a+b)+(c+d), which can be processed in parallel, with two processors each contains one stack and only needs 2 time steps (unit of addition). The worst is abcd+++, or a+(b+(c+d) ), which needs 3 stacks and takes 3 time steps. The remaining ones, abc++d+ and abc+d++, are in between the two extremes. The figure below exhausts all the five trees corresponding to the algorithms to add four numbers a, b, c, and d.

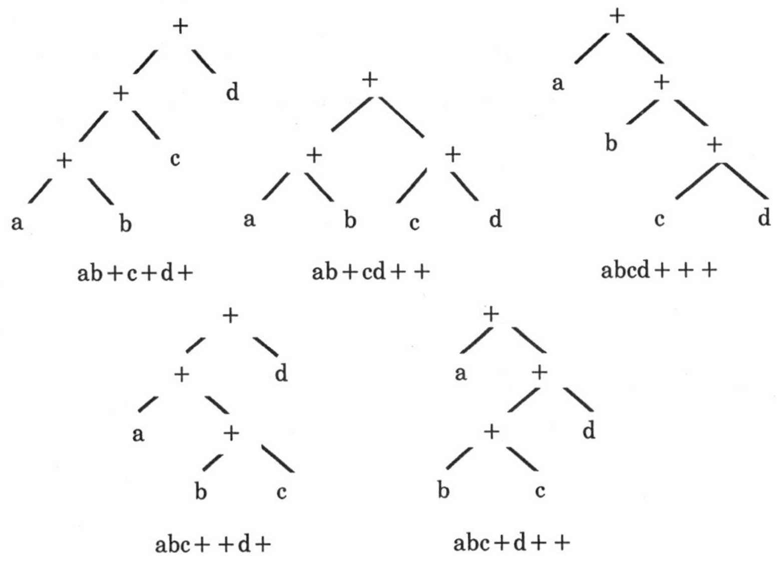


Fig.1

The numbers of different trees $f_n$ for fixed number of leaves $n$ are able to be enumerated by carefully exhausting possible diagrams, and are tabulated below.

| $n$ | 1 | 2 | 3 | 4 | 5 | 6 | 7 | 8 | 9 | 10 | ... |
|---|---|---|---|---|---|---|---|---|---|---|---|
| $f_n$ | 1 | 1 | 2 | 5 | 14 | 42 | 132 | 429 | 1430 | 4862 | ... |

It will be an interesting puzzle to find an expression for the n-th member of this series. The solution to this puzzle will be given in the next section.

## 3. Explicit expression for the number of binary trees

We firstly demonstrate the method of deriving a closed form concerning general progressions. The well-known Fibonacci series is employed as an example. The n-th Fibonacci number $f_n$ is defined by the sum of its two predecessors:

$$f_n = f_{n-1} + f_{n-2} \tag{1}$$

with the initial values

$$f_0 = 0 \text{ and } f_1 = 1. \tag{2}$$

We define the generating function of the series as,

$$F(z) = \sum_{n=0}^{\infty} f_n z^n. \tag{3}$$

Multiplying $F(z)$ by $z(1+z)$ and comparing it with $F(z)$, we have

$$z(1+z)F(z) = -f_1 z + f_0(z-1) + F(z). \tag{4}$$

Using the initial values, we obtain

$$F(z) = \frac{z}{1-z-z^2}. \tag{5}$$

$F(z)$ is expanded as,

$$F(z)=\frac{z}{p-q}\left(\frac{1}{p-z}-\frac{1}{q-z}\right), \tag{6}$$

where p and q are the solutions of the quadrutic equation

$$z^2+z-1=0, \tag{7}$$

that is,

$$p=\frac{-1+\sqrt{5}}{2}\ and\ q=\frac{-1-\sqrt{5}}{2} \tag{8}$$

Equation (6) is expanded as a Taylor series, and we finally obtain the well-known expression

$$f_n=\frac{1}{\sqrt{5}}\left(p^n-q^n\right). \tag{9}$$

More generally, $f_n$ is given as a function of $f_0$ and $f_1$

$$f_n=\frac{1}{\sqrt{5}}\left\{f_0\left(p^{n+1}-q^{n+1}\right)+(f_1-f_0)\left(p^n-q^n\right)\right\}. \tag{10}$$

Similar method can be applied to count the number of binary trees. Taking off the right most binomial operator in the inverse Polish terminology, one expression is decomposed into two independent parts; one with $i$ terms and the other with $n$-$i$ terms.Considering this decomposition for all possible $i$'s, we find the recursion relation

$$f_n=\sum_{i=1}^{n-1} f_i f_{n-i}. \tag{11}$$

Let's introduce the generating function again,

$$F(z)=\sum_{n=1}^{\infty} f_n z^n. \tag{12}$$

According to the form of Eq. (9), we make a square of $F(z)$, and easily find that

$$[F(z)]^2=F(z)-f_1 z, \tag{13}$$

with the initial condition $f_1=1$.

From Eq.(13) we have

$$F(z)=\frac{1\pm\sqrt{1-4f_1 z}}{2}. \tag{14}$$

We must take the minus sign to get the correct sign for $f_1$. This expression is expanded in Taylor series, and finally we obtain the closed form of $f_n$ as,

$$f_n = \frac{(2n-3)!!2^{n-1}}{n!}$$

$$= \frac{2(2n-3)!}{(n-2)!n!} . \qquad (15)$$

For very large $n$, Stirling's formula gives

$$f_n \sim 2^{2n}, \qquad (16)$$

and

$$log_2 f_n \sim 2n . \qquad (17)$$

This leads to the conclusion that we consume at least about two bits of information per word to select the desired algorithm out of all possible ones. Theoretically, the optimum efficiency of the mapping program to PLU is limited by this value. In other words, the goal for the mapping programs is this value.

## 4. Part of Speech

Tree diagram is often used in syntax analysis of sentences. Especially, the immediate consituent (IC) analysis[3)] uses mainly binary trees. Generally speaking, IC analysis has been successful in this field. While reading a sentence, if one finds that the sentence makes sense, it means that one has already implicitly recognized one of the tree diagrams that makes the series of words gramatically valid. For example, ' A young man with a paper followed the girl with a blue dress' becomes meaningful when we assume the IC structure;

*(((A)((young)(man)))((with)((a)(paper))))*

*(((follow)(ed))(((the)(girl))((with)((a)*

*((blue)(dress))))))* . (18)

This IC structure is expressed by the binary tree diagram in Fig.2.

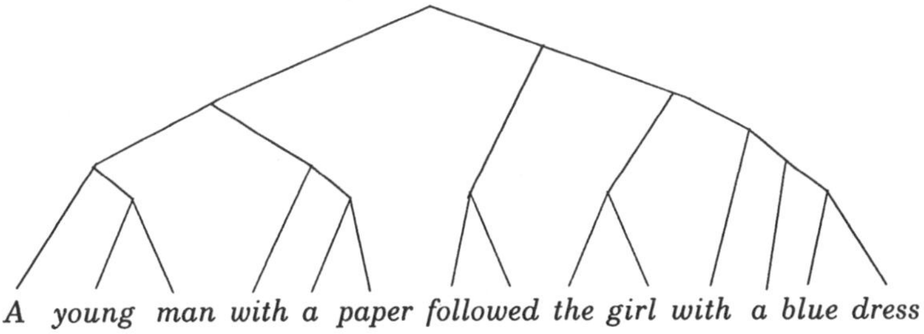


Fig.2

The amount of information (2 bits per word) found in the previous section has the meaning that there should be more than or equal to four grammatical cathegories of word (part of speech) so that one can resolve the ambiguity of the IC structrure of any given sentences. In other words, to be able to understand long sentences written in a language, one should have learned at least that amount (two bits per word) about the language. This may provide a reason that infants can handle with sentences consist of only one or two words, for which there are no ambiguities concerning the IC structure.

It is well known that there are some examples for which IC analysis fails to resolve the ambiguity in the meaning. A well-known example against IC was found in the joke; '"Time flies. " "You can't; They fly too fast. "'[3] This example will suggest that the nessessary number of part of speech will be more than 4. In the counterevidence mentioned above, the first speaker regards "time "as a noun and " fly" as a verb, but the second speaker does conversely. A close consideraton of this case leads to the conclusion that two sentenses with the same IC structure must be classified at least into two classes; for example, S+V and V+S. We must take account of this ordered structrure. This will be done by introducing two kinds of binomial operators in the terminology of inverse Polish notation. Inclusion of this ordered coupling of two elements in a sentense leads to the number of trees in Eqs. (14) and (15) multiplied by a factor $2^{n-1}$. The same result is also obtained by multiplying the right hand side of Eqs. (10) with the initial condition $f_1 = 1$. In this model the number of parts of speech must exceed 8. It should be noted that sentenses of two words and three words both require the same number 2 for grammatical classes. We guess that an infant that understands two-word sentenses can also understand three-word ones, eventhough it can speak only two-word sentecses, that is, it can decompose three-word sentenses but cannot construct them. There will be more complexities in the composition processes than in the decomposition processes. To ask what the difference between the two processes is is beyond the scope of the present paper.

We are very grateful for Miss Kumie Suzuki for her beautiful operation of word processing of this manuscript by J Star,which is offered to Education Center for Information Processing, Tohoku University by **FUJI XEROX Co.**.

## References

1) "Gödel, Escher, Bach" , D. R.Hofstadter, (Basic Book Inc., NY, 1979).

2 "Programmable Logic Unit" , Ph.D Thesis, Tohoku University, H. Yamada, 1985.

3) 'Grammer' , H. Palmer, (Penguin Books Ltd., Harmondsworth, England, 1971).

"Understanding Natural Languages" , T . Winograd, (Academic Press, 1972, NY).

AUTHOR INDEX